XPONENTIAL 2023

AA006161

Denver, Colorado, USA
8 - 11 May 2023

ISBN: 978-1-7138-9932-7

Printed from e-media with permission by:

Curran Associates, Inc.
57 Morehouse Lane
Red Hook, NY 12571

Some format issues inherent in the e-media version may also appear in this print version.

Printed by Curran Associates, Inc. (2024)

Association for Unmanned Vehicle Systems International
2700 South Quincy Street
Suite 400
Arlington, VA 22206
USA

Phone: (703) 845-9671
Fax: (703) 845-9679

www.auvsi.org

Additional copies of this publication are available from:

Curran Associates, Inc.
57 Morehouse Lane
Red Hook, NY 12571 USA
Phone: 845-758-0400
Fax: 845-758-2633
Email: curran@proceedings.com
Web: www.proceedings.com

TABLE OF CONTENTS

AI FOR DRONE AUTONOMY

Stefan Milz[1]

The paper gives an overview of current challenges for AI in the area of safety-critical scenarios for real-time drone and robot automation. Exemplarily this is investigated for different use cases,, e.g. Ground-risk minimization in the case of Precise Automated Landing for Delivery of goods. A methodology is presented on how a system with AI needs to be designed using operational domain definition, scenario based analytics, requirement traceability throughout neural networks and safety by design. These principles are generically applicable also in automotive or other fields. The presentation will open eyes for a new development of AI using large scale data models and motion based functions. This paves the way for integrating AI into the certification process.

INTRODUCTION

Partial Automated drone flights in the form of data service, but also package delivery as well as security applications has more than tripled in recent years. However, the road to full autonomy is still long. Above all, the real-time analysis of data, including AI, has not yet been fully developed. The certification of such systems has also not yet been fully thought through.

A combination of different sensors - e.g., cameras, laser scanners or radars - is necessary for reliable perception of the environment in different weather conditions and ambient conditions. Especially for semantic mapping of the environment, i.e. the recognition of objects or persons based on sensor data, classical rule-based methods are limited because deterministic methods are not real-time capable due to the large amount of data.

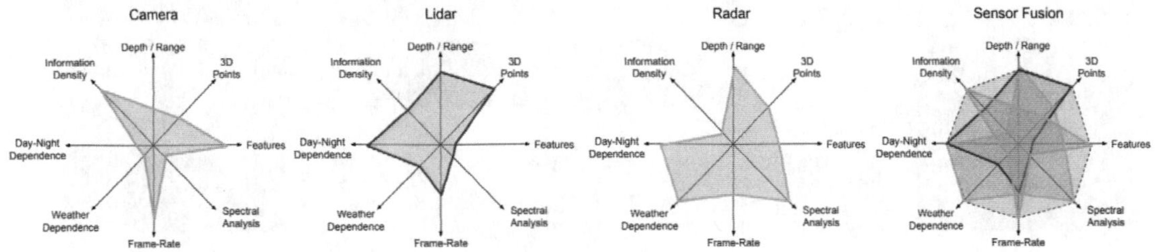

Figure 1. Different Sensors have different strengths and weaknesses for Drone Autonomy

Artificial intelligence (AI) approaches, in particular machine learning (ML), have the potential to remedy this situation, as they can be highly parallelized and, despite their high performance and statistical properties, also enable a large data throughput and thus ensure real-time capability.

[1]CEO, Spleenlab GmbH

However, the use of AI methods in safety-critical systems of aircraft has not been possible so far for reasons of verification and certification. This is mainly due to their statistical nature, which makes a behavior of an ML algorithm unpredictable. However, this predictability, also called determinism, is one of the most important building blocks in the certification process of safety-critical systems. Currently, there is no approved approach on how a proof of concept to ensure safety for ML algorithms should look like. With current methods, almost infinite amounts of data are required for verification.Autonomous flight is a rapidly evolving application area with many opportunities for AI or machine learning based approaches. In general, two different pipelines can be distinguish:

1. the mediated perception approach, which semantically reasons about the scene and based on that determines the flight control decision.

2. the end-to-end approach, which determines flight control based on human behavior in an end-to-end approach.

Fig. 2 provides an overview of the two pipelines, showing examples of possible applications. (a) SLAM is crucial for the local map and vehicle position within the environment model. (b) Scene Understanding is crucial for the interpretation of the environment, e.g., for the detection of static and dynamic objects and their locations, such as point-based classifications. (c) Sensor fusion is important for exploiting the strengths of different sensor types, e.g., classification for cameras, reconstruction for lidar, or Dynamics for radar. (d) End-2-end flying learns all decisions within a single network and can be treated as an alternative approach. Compared to other types of automated vehicles, autonomous flying has specific challenges that characterize AI use cases:

- Scale ambiguity: the 6DoF vision of aerial vehicles complicates basic geometric geometric tasks such as visual depth estimation or visual reconstruction compared to 3DoF use cases for ground robots and cars.
- Data availability: Public datasets are scarce compared to other computer vision tasks.
- Constraint Hardware: Applications have to run on a limited hardware with low energy consumption.

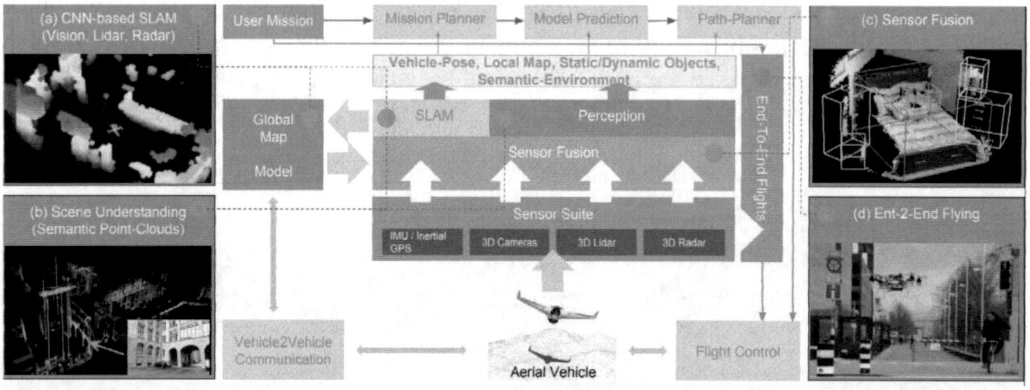

Figure 2. The principles of automated flying: The diagram outlines the state-of-the-art workflow. It all starts with the dedicated user mission. However, vehicle sensor data is essential to develop an environmental model for decision making and path planning. Several sensors like cameras, Lidars or Radars are crucial. In general two different paths are distinguished: 1. The mediated perception approach; 2. End-2-End Flying; Due to the complexity of the different tasks, leading approaches

are mainly based on Machine Learning or Deep Learning. (a-d) Illustrate example functions based on Deep Learning and their specific role within the pipeline.[2]

AI CHALLENGES FOR DRONE AUTONOMY

Classical approaches of the development of systems with an increased safety risk, for example autonomous drones, typically follow a W-model. Based on a risk analysis, system requirements and successive hardware and software requirements are defined. These are implemented in a waterfall process and individually tested on software and system level in simulation as well as in the real world, in scenario-based testing. This is very straightforward. It classically requires a high degree of determinism, from perception to control, which is inherently not present in a neural network. Therefore, the classical approach needs to be reconsidered and combined with the features of modern AI. EASA[3] (see Fig. 3) has presented such a development pipeline for the first time in its concept paper. The target is a deterministic process development chain throughout a neural network.

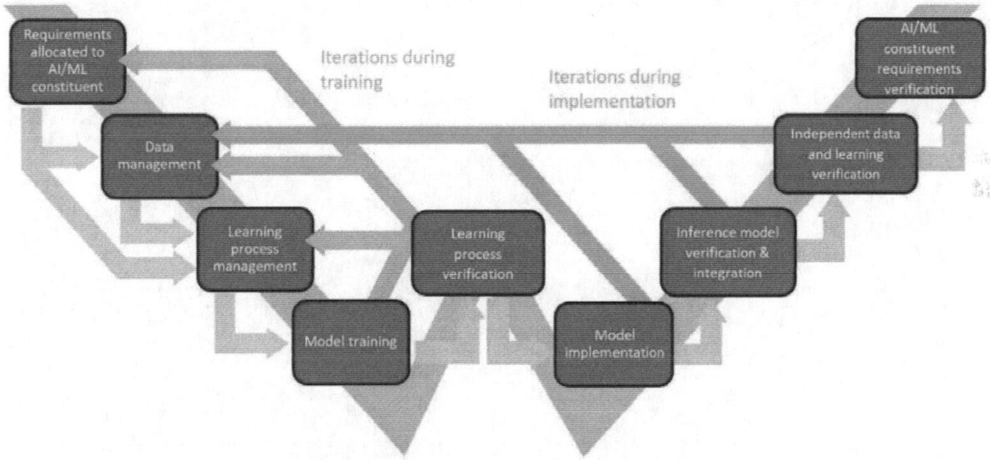

Figure 3. W-Modell update for AI by EASA Concept Paper

OPERATIONAL DESIGN DOMAIN & USE CASE DEFINITION

For each autonomy use case, especially in the drone domain, the operational design boundaries must be defined. This means that the system requirements and their significance for the sensor technology must be derived directly from the intended use. This is the most common mistake of modern systems, since sensor technology is defined and frozen before the actual use case and the accompanying Operational Design Domain (ODD) is defined. Typical use cases for drones in modern automation of drone operation are:

A. Vision based AirTraffic recognition for **Detect & Avoid** (AirRisk Estimation)
B. Vision based Near Field **Collision & Obstacle Avoidance**
C. Vision based Navigation for **Safe Precision Landing** (Ground Risk Estimation)

---

[2] Milz, et. al, "AI-Survey for Self-Flying Vehicles" Nonlinear Dynamics. NIPS 2018 Workshop MLITS

[3] EASA et al. "First usable guidance for Level 1 machine learning applications", Link: https://www.easa.europa.eu/en/newsroom-and-events/news/easa-releases-its-concept-paper-first-usable-guidance-level-1-machine-0

D. Vision based **GPS denied operation** for semi & full Drone Autonomy

Some impressions of these use cases are shown in Fig.4.

Figure 4. Four sample Use case for Drone Autonomy: A- Detect & Avoid; B- Collision & Obstacle Avoidance; C - Safe Precision Landing; D - GPS Denied Operation

it is necessary to precisely specify the framework of the use case to be operated in a so-called ODD. The ODD will directly have an impact on specific development properties, e.g. the data set for an AI algorithm. The impact of the ODD influencing systemic AI artifacts is shown in Fig. 4

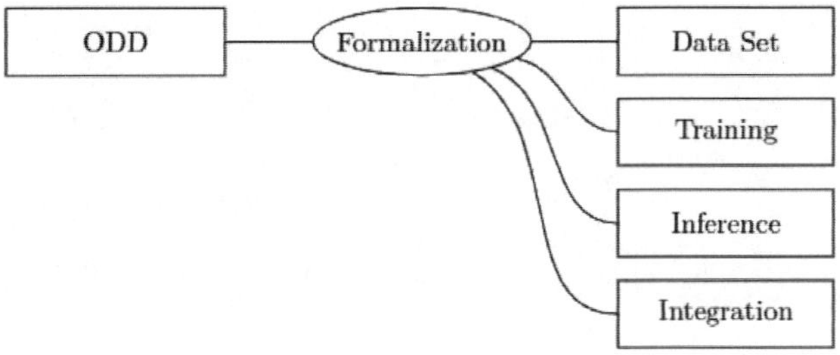

Figure 5. Impact of ODD with AI based Processes

A well-defined ODD has boundary conditions that specify implicit sensor properties and AI performance. The more precisely the ODD is defined, the better the requirement for AI and sensory perception. An example ODD for different use cases, e.g. Detect & Avoid such as Precision Landing is shown in Figure 6.

Attribute	**Value**
Flight height	0 m–120 m
Velocity	0 m/s–10 m/s
Expected intruders	helicopters, paragliders, UASs
Light	$\geq$ 1000 lx
Precipitation	none
Wind	$\leq$ 5 m/s
Temperature	5°C–40°C
Area	mountainous, hilly, plain

Figure 6. Sample ODD Description

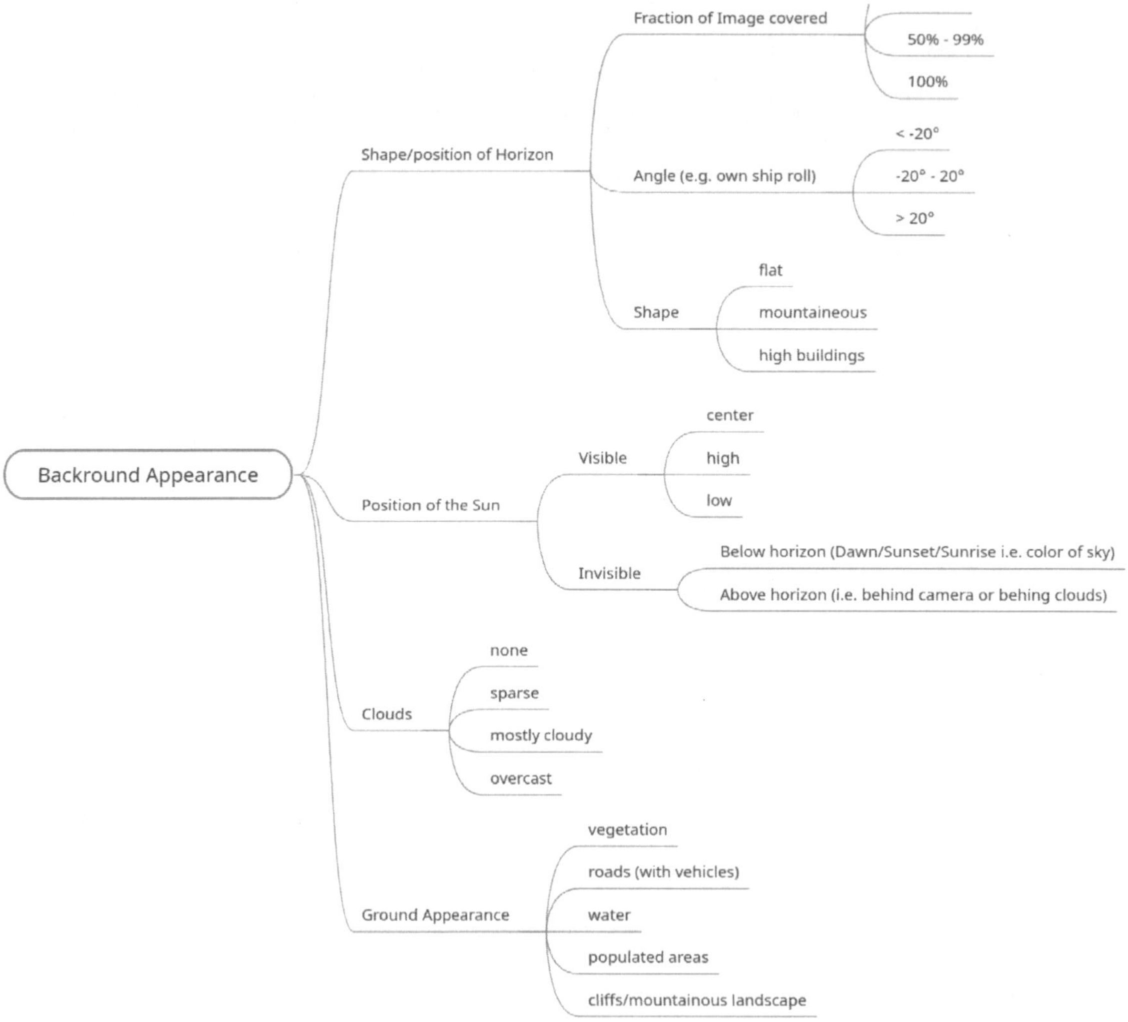

Figure 7. Sample Divergence of a data set for AI derived from well-specified ODD

The divergence of a data set for AI could be derived from a well-specified ODD. Fig. 7 is an exemplary divergence tree solely for the background appearance artifact of camera data in the ODD. Such a divergence analysis needs to be designed for all artifacts of an ODD. In Fig.3 is the development model explained. All starts with the requirements. Those are typically derived from the ODD. Each requirement artifact needs to be analyzed in its divergence and for each artifact could be drawn in a divergence tree. For all classes in the tree needs to be test, training and verification data. Such availability could be drawn in an availability matrix to see if there is sufficient data for each artifact. If classes are not occupied, requirements are not served and the system will be trained and validated insufficiently. Fig. 8 shows such a matrix, where colors identify specific scenes that are present in the data. A single cell is a divergence class. White cells are missing scenes. Hence, these classes and therefore the artifacts are not present in the data.

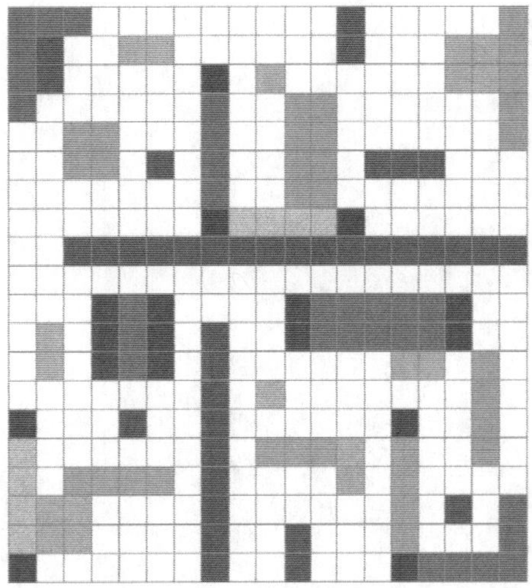

Figure 8. Availability matrix for Data

SAFETY BY DESIGN USING THE EXAMPLE OF PRECISION LANDING

A redundant system design is necessary for precise and safe landing. By means of a modular approach, a safe observation of the ground risk during the landing can be performed on the basis of several sensors. Fig. 9 shows exemplary redundant Design for such a Landing strategy.

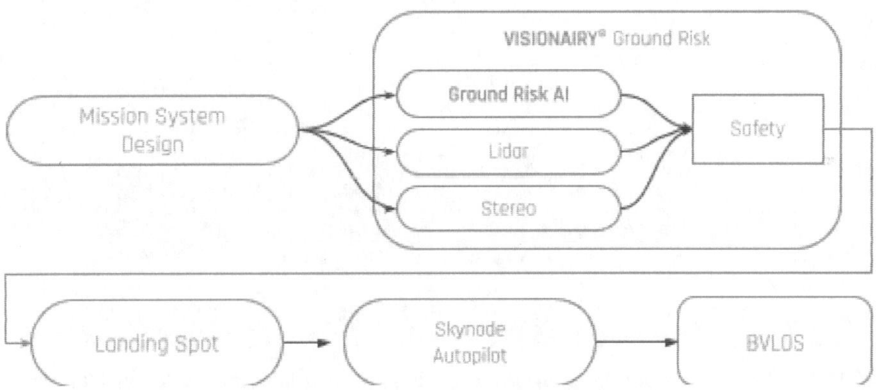

Figure 9. Safe System Design for Landing

Vision based Precise landing of a drone is a scientifically solved problem. Real operations, especially in interaction with humans, for example in the field of parcel delivery, is currently an unsolved problem due to safety regulations. A Safety By Design approach, modular with redundant sensors, can be used to detect people or vehicles during landing by means of an RGB camera (Fig. 10) or obstacles and inclination by means of a depth sensor (Fig. 11). This supports safety during the landing process.

Figure 10. RGB based Object Detection during Landing in Simulation - Collision Detected (up) and Emergency Landing (down)

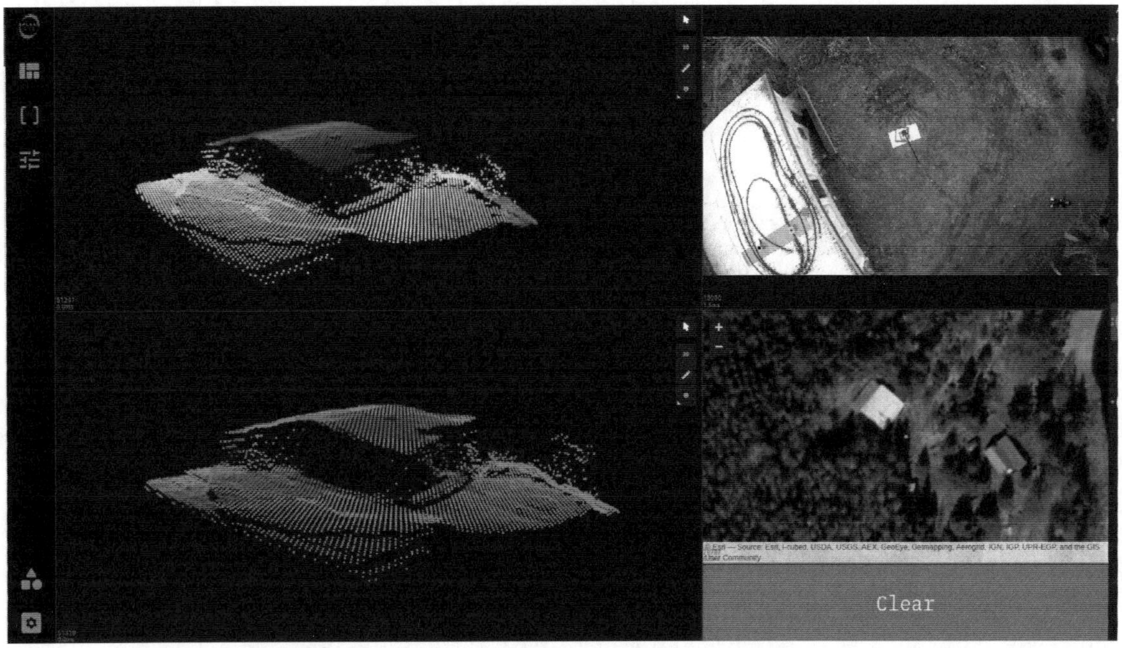

Figure 11. Depth Sensing based Collision Avoidance during AI based Landing

CONCLUSION

Integrating AI into real-time systems like drones is extremely challenging due to safety requirements. We have highlighted the basic problems and shown methods such as traceability, data set mapping, ODD design, and Safety by design. Furthermore, we have shown use-cases where especially AI perception includes a significant improvement to the operation of drones: Detect & Avoid, GPS-Denied, Collision Avoidance and Safe Precise Landing. One System Design for integrating AI was demonstrated on the landing use case with the aid of System by Design. There are still a number of challenges, but the first foundations have been laid.

AUTONOMOUS AIR-TO-AIR REFUELING SYSTEM AND TETHERED MACHINE VISION GUIDANCE (A3R®)

Marinus B. (Ben) Bosma*[1]

This paper describes a new and novel method and apparatus for connecting two airborne bodies in flight. Specifically, the air-to-air refueling of unmanned aircraft but can be scaled to any flying or mobile platform requiring physical connection while in motion. I also provide a means of developing this system using ground-based testing as well as airborne flight testing.

INTRODUCTION

Modern Technology Solutions, Inc. (MTSI) is an employee-owned engineering services and technology solutions company delivering first-choice capabilities to solve problems of global importance. Our mission is to make important and lasting contributions to the nation's defense and security by providing leadership and best-value solutions to solve America's most technically challenging strategic problems. This paper describes the lynchpin of MTSI's patent portfolio of fixed-wing, tethered applications.

A3R® is an Autonomous Air-to-Air Refueling system that makes the receiving aircraft passive. This technology has the potential to vastly improve the overall safety of aerial refueling missions by accurately maneuvering the hose to the receiving aircraft. A3R® is not dependent on the traditional probe and drogue technology and allows a crewed or uncrewed tanker to refuel autonomous systems in a manner that keeps them focused on the mission at hand.

A3R® is.

- Platform agnostic
- Multiple configurations
- No extensive aircraft mods needed.
- No boom operator required.
- Increased safety for operations
- Significantly extends combat radius.
- Keeps autonomous systems focused on the mission and not flying to the tanker

Aerial Refueling

Refueling aircraft in flight is a well-known process, and there are many examples of systems and equipment in the art provided to accomplish the refueling of aircraft. All such systems require automated or user-directed cooperation of an aircraft being refueled to accomplish the process. For example, in many such systems, a tanker aircraft positioned proximate to an aircraft to be refueled and trails a hose for transferring fuel. Typically, the hose

[1] Principal Engineer, Modern Technology Solutions, Inc. (937) 478-6705, ben.bosma@mtsi-va.com

Figure 1. Early Air to Air Refueling 1928

is dry, meaning that there is no fuel in the hose at the time of connection of the hose to a receiver at the aircraft to be refueled. Personnel on board the receiving aircraft are then responsible for maneuvering the receiving aircraft into a position where the fueling hose may be captured and connected to receiving equipment. There are many problems associated with this process, not the least of which is, that the receiving aircraft must be diverted from whatever mission it may be conducted to accomplish the refueling process, and then must redirect back to the mission at hand.

What is clearly needed is a system and process in which the receiving aircraft may be refueled without interruption, and transparently to the receiving aircraft, its operating systems, or its operating personnel.

Figure 2, The Author Flying Chase Aircraft During MV-22 Refueling Testing

REVERSING THE ROLES

There have been numerous systems for autonomous air refueling that mainly focused on re-purposing the existing body of air-refueling hardware namely probe and drogue whereby a hose is trailed behind a tanker aircraft with a high-drag drogue. The receiving aircraft has a probe that enters a pocket in the drogue and mechanically latches and seals to the drogue. This process requires the receiver to maneuver precisely and not without considerable force to engage the drogue to make proper contact. The second method is the boom and receptacle where a flyable boom trails behind a tanker and by some lateral and mostly axial extension inserts a nozzle into a fixed receptacle on the receiving aircraft. In both methods, the receiver is required to maneuver into a precise and fixed location to make contact with the fueling apparatus.

The A3R® approach does not require the receiver and tanker to maneuver into precise positions but instead allows them to be loosely positioned by leveraging A3R®'s technology. This technology includes a flexible, slender, and relatively long conveying hose which automatically maneuvers itself via an adaptive motion control system to connect into a receiving aircraft receptacle to then pass fuel (or power) to it.

This method whereby the hose becomes a robot arm allows considerable flexibility both in scale and maneuverability. Scale, in that, this method can just as easily be a large hose capable of dispensing huge volumes of material or a slender hose hundreds of meters long and able to find, perch on and refuel light, Class I UAVs.

MODELING AND SIMULATION

Early in the development of the concept of A3R®, there was skepticism about whether a small flying body can deliver a hose to a fitting on the surface of a wing. Various comments such as “bow wave effect”, “complex flow over a wing”, “suction force bashing the end effector onto the top of the wing”, “sloshing of fuel in the hose” and more were voiced over the concept. Even the drape of the hose with the end effector on the end was questioned in my graphic depictions. The only thing to do was to do some computational fluid dynamic tests to see how bad it really is.

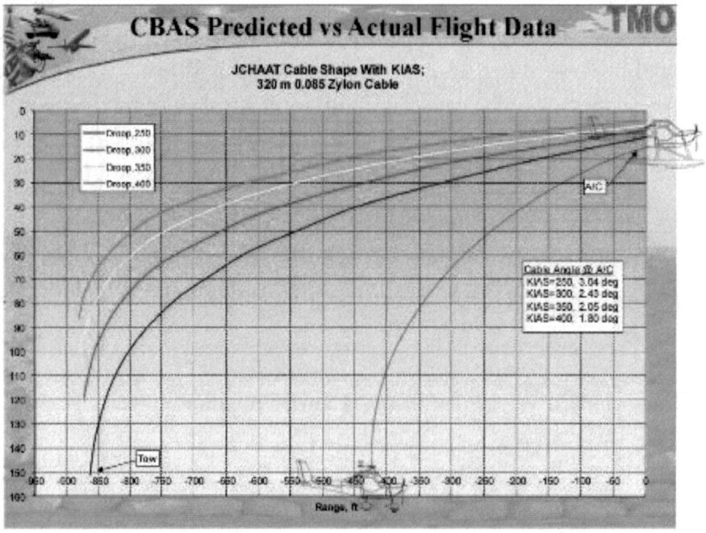

Figure 3, Cable Drape on Thin Cables (Army)

The question about whether a hose would hang down is difficult to answer. I tested several CFD approaches none of which had easy solutions. The problem with CFD on an elastic (non-linear) body is that each iteration is several hours of computation because the hose at even 50ft in length is very dynamic. It generates aerodynamic flow like a cylinder when it is vertical but as it curves into a stable shape the angle the relative wind takes to the hose changes from a very elliptical and stable shape to a vertical and more cylindrical shape. As this continuous curve develops the airflow and forces on the hose change proportionally. The recursion in solving for the physical shape of the hose in flight required hundreds of iterations and therefore hundreds of hours on a fast PC. So, I looked at the history of air-refueling and made some assumptions. In Fig 1 you can see how a standard fuel hose behaves when deployed from a slow-flying tanker. The hose drapes down. Fig 3 is a cable drape chart from Army testing for a 320m cable. Our own testing confirms that a hose charged with fuel goes almost straight down.

The aerodynamics of the end effector over a wing was a little more manageable because there is little elasticity in the bodies. I am using a wing section based on the Merlin PSA which I’ll get to later.

Fig 4 is the Merlin wing with the fueling blade and no hose in the Computational Domain. I'm limited in computer power and skill so bear with me. I'm using the Merlin as the target aircraft for both tanker and receiver because it is a very good analog to Class II UAS. The left side is 36" from the centerline and the right side is 108" from the centerline. There is still another 6ft of the wing but I'm working with limited resources so I'm trimming the center 6ft out for testing. The actual fuel filler cap is within 1/2" of the center of that wing panel. I put the refueling blade in that hole.

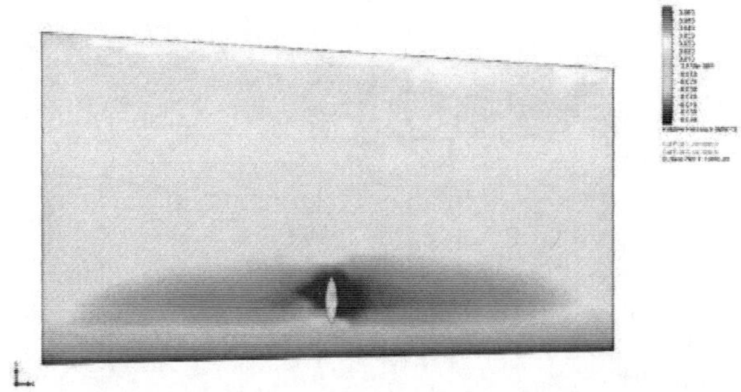

Figure 4. CFD Model with Fueling Blade Top View

We're seeing some rollover flow from the tips, some difference in bow pressure due to taper and maybe a little from the 2deg washout. There is more lift on the outboard side of the blade than the inboard due to the span wise flow and the blade acting like a fence. I want to add that the wing model is based on a spline I copied from the sheet metal model of the Merlin. It has about 8 control points which are way too many to make a smooth curve. If I had the parametric equation for this wing, I'd be a lot happier. I just did a best fit to the model since the sheet metal model was discontinuous. The airfoil is the same from root to tip, but it tapers and twists down 2deg. There is also a 3deg dihedral which is the same as the target asset, however, the target asset wing has no washout and is a constant cord most of

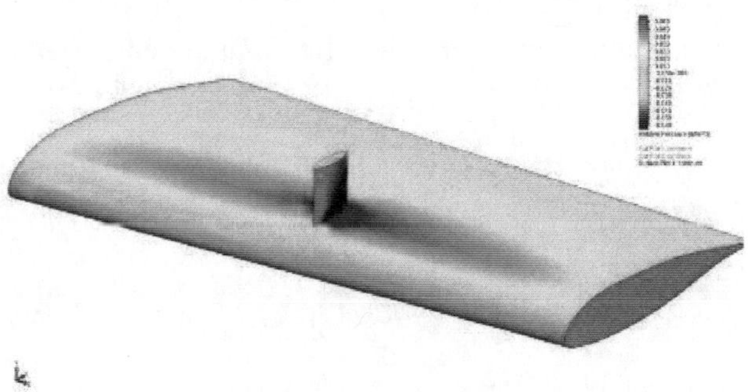

Figure 5. CFD Model with Fueling Blade Oblique View

the way out.

The target asset wing doesn't have the wing loading the Merlin has. We can safely assume it will be an easier wing to perch on.

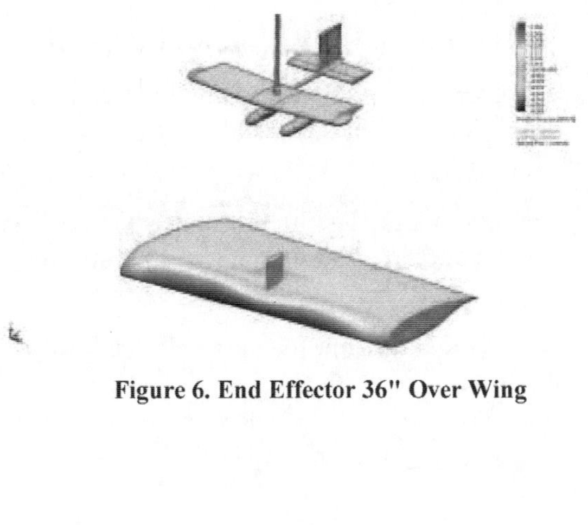

Figure 6. End Effector 36" Over Wing

Here is the end effector at 36" from the blade. There are actually a few hundredths of a PSI of pressure increase on the wing. I can't explain that.

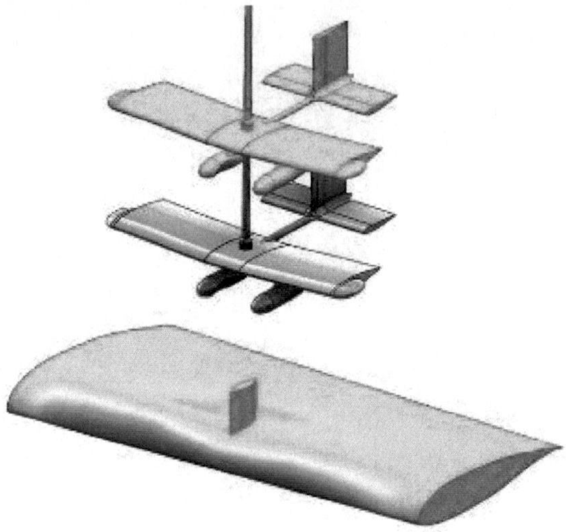

Figure 7. End Effector 18" from Wing

If I drop the end effector down into the wing 18" now, we can see a noticeable increase in pressure on the front of the wing. It's still very slight only .005psi but it's there. My mesh is very loose because I don't have a lot of CPU to go through these cases, quickly. I'm just trying to get a sanity check. FYI (150,000 nodes in my mesh, very rough compared to the 15 million nodes in a rubber hose model)

What this shows is that there is a disturbance in the flow field but the end effector itself isn't affected much from 36" down to 18" from the blade.

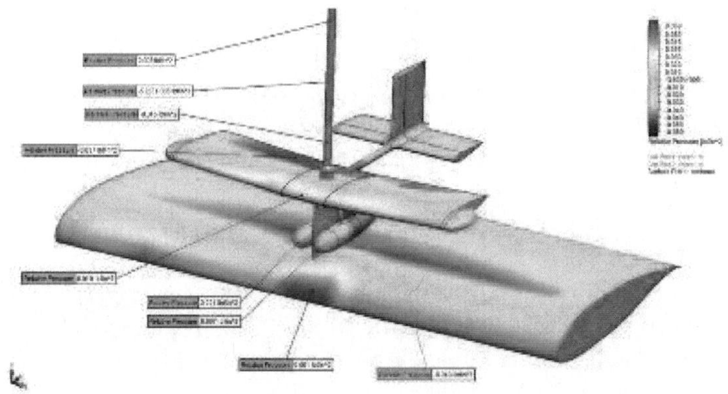

Figure 8. End Effector Perched on Refueling Blade

With the end effector on the Blade, there is a pressure buildup on the end effector. That means we need to crack the throttle. BTW, this is nothing a fighter pilot doesn't already know when he tucks himself into the slot. It takes power to get into a tight formation.

Finally, the end effector is clamped to the wing. The fuel hose has always been charged with fuel and I don't really know how to address the "sloshing" concern other than it's beyond my understanding that fuel will slosh in a fully charged hose.

Fig 5 shows the velocity flow lines looking at it from the side. Notice how much difference there is in velocity. The freestream velocity is 60knots.

The bottom line is that the flow over the wing is very smooth and doesn't bother either the host or the end effector, itself.

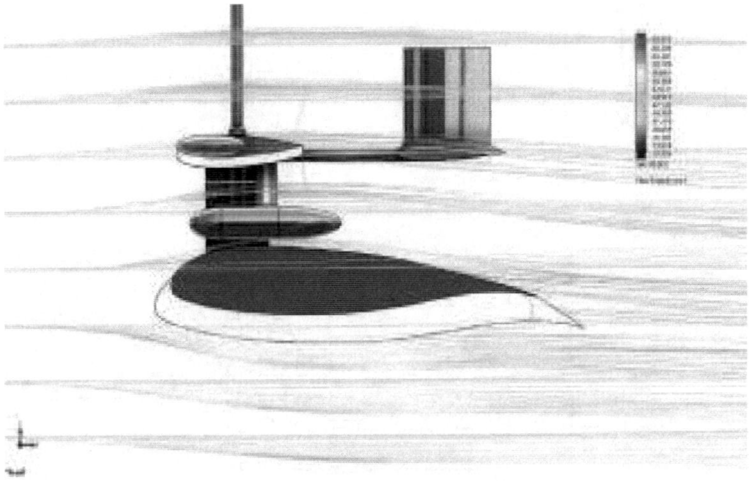

Figure 9. Flow Over Wing with End Effector Perched

Ground Testing

So much for CFD: Time to put some air over the system.

For that, I designed a test using a truck, trailer, and sailboat mast. I hinted that we're going to look at the Merlin for flight testing. We're not going to mount Merlins to a moving truck, but we can simulate the exact geometry of a flight test with this setup.

Figure 10. Ground Testing Concept

The ground test will replicate what we'll see in the air. That way we can pre-qualify the A3R®, software, and CONOPS before we commit to a flight test program.

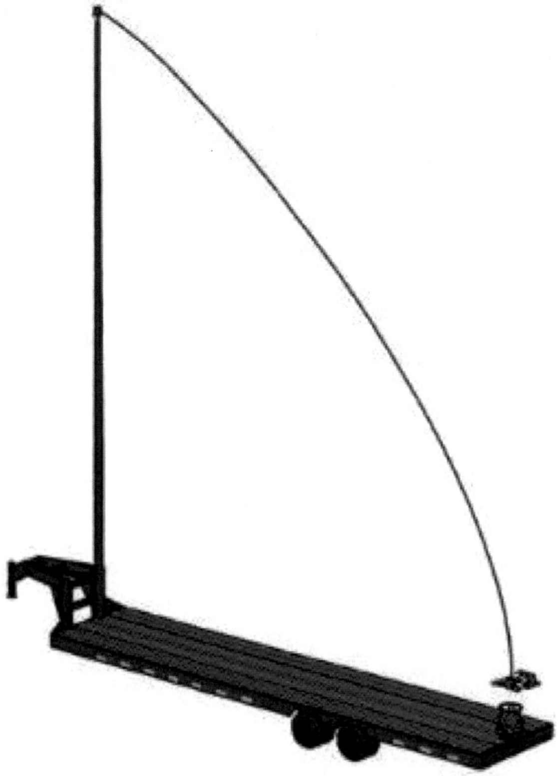

Figure 11. Trailer Test Rig

The first step is to understand the behavior of the end effector on a tethered hose. For this, we use a sailing mast. The sailing mast is an aluminum extrusion with all the necessary features for hoisting the hose in the air, adjusting its length and vertical anchor position. We can map the location of the end of the hose as a function of airspeed. This gives us the relative positions of the two aircraft so we can ensure a proper formation to execute a contact. We can also try different hoses, end effectors, and airspeeds up to 126knots on the test track. Having a ground-based testing platform like this will greatly increase the chances of success in the air.

Flight Testing

I propose testing A3R® with crewed aircraft and specifically the Merlin Personal Sport Aircraft (PSA). The Class 2/3 UAV community falls in the middle of the Merlin's flight envelope.

This is the Merlin PSA compared to a known Class 3 UAS we built for a defense customer. This platform has over 50,000 combat hours and is still in service but is limited to only 30hrs of flight time. Being able to refuel this asset in the air would be very beneficial.

Here is what the relative sizes are.

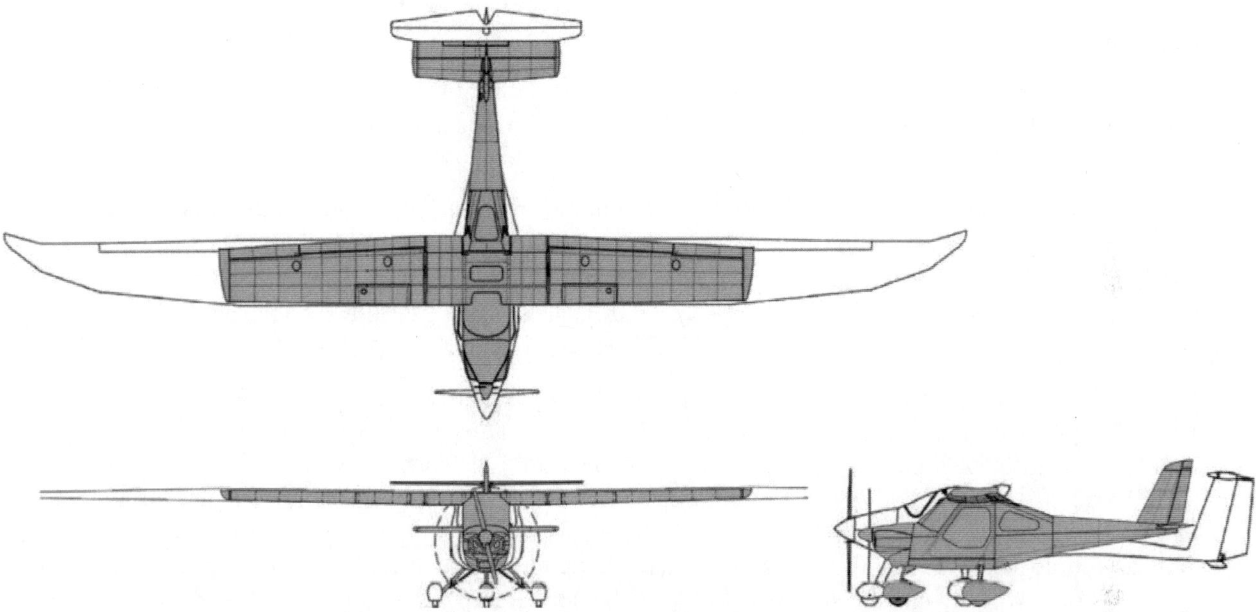

Figure 12. Size Comparison Between Pipistrel Sinus and Merlin PSA

The Merlin has a 6 ft. center section that takes the wing attach points out farther than that of the Pipistrel. This makes the wings very light. Pipistrel wings weigh 145lbs each empty. Merlin wings weigh 38 lbs. each. The Merlin can be assembled by 2 persons. The Pipistrel requires a minimum of 3.

The Merlin has the same tires as the Pipistrel but has substantially more ground clearance than the Pipistrel because it swings a smaller prop.

REGULATORY CONSIDERATIONS FOR FLIGHT

Two issues come up when putting things in the national airspace (NAS) for flight testing.

1. The aircraft
2. Where you fly the aircraft.

We want to demonstrate the capability of one UAV refueling another UAV, autonomously. The FAA has something to say about it, but ATC is okay with it if the aircraft are squawking and talking. Both these activities are easily done with a transponder and repeater radio. ATC does require that the person talking be a pilot no matter where they are at a ground control station or in the cockpit.

The manufacturing of the aircraft is another issue that the FAA has something to say about namely because it keeps a lot of folks employed inspecting, writing specs, and telling engineers how to do their jobs. The notable and exploitative exception to that is the EXPERIMENTAL aircraft category.

Since this country values invention and innovation, there is still an EXPERIMENTAL category on the books. This category was created so private individuals and companies can design, build, and test new aircraft and aircraft systems.

There are several subcategories of EXPERIMENTAL, but the sweet spot is the **EXPERIMENTAL-amateur built** category which is commonly called a homebuilt. This is the space where the aircraft is built by the owner/operator. For an airplane to qualify for this category the owner must do 51% of the work. There is no negotiating this, so the FAA specifically names the kits that qualify. The 51% rule is so third parties don't pop up making airplanes for people. It's all designed to maintain the integrity of the category.

If we modify a certificated aircraft we run into a host of problems, namely getting approval from the manufacturer, Flight Standards District Office (FSDO) who are the same as DCMA for aircraft builders and people who type because you'll have to provide reams of paperwork with specific persons doing the tasks to submit to the various regulating bodies. If we modify an EXPERIMENTAL, we only need to answer to ourselves with a notification to the FSDO about what we're going to do. There are "minor" and "major" modifications so it's not quite as simple as it sounds but it's still easier than changing an airworthiness certificate on a certificated aircraft. Some people will say that the solution to the certificated aircraft limitations is to slap an EXPERIMENTAL decal on the side and be done with it. It's doable but significantly more difficult because the FAA takes a dim view of people modifying aircraft and pretty much guarantees that that aircraft is never restored back to its original airworthiness certificate.

The other issue we have is where to fly a UAV. Since we're looking at something in the same class as the LEAP, we probably can't fly it outside of restricted airspace. Most UAV waivers deal with small spaces and altitudes below 400ft. We want to operate up to 15,000ft and BLOS. That's a challenge to be approved for an unmanned platform. The answer is to *put a pilot onboard.* The FAA, ATC, and airport operators are very comfortable with pilots flying airplanes in and out of their airspace.

EXPERIMENTAL aircraft fly almost everywhere in the NAS if they are equipped to go there, and the pilot flying is qualified. For example, to fly into class B or C airspace, you need a transponder that can squawk altitude and position. For class B you need an IFR clearance which also implies special equipment and pilot qualifications.

So how do we overcome this? Again, *Fly an EXPERIMENTAL aircraft with a pilot onboard*. The experiment is to fly unmanned but that only means we provide the platform with all the functionality of a UAV without removing the capability of the safety pilot assuming full control of the aircraft. That's easy.

There are a wide variety of aircraft kits on the market, but none offer the kind of benefits of the Merlin PSA over a Pipistrel Sinus, for example.

The Merlin is a new generation of quick-build homebuilt aircraft that follows a philosophy of functionality and utility. It is a single-place, high-wing, all-metal ship of very conservative construction.

Here are some comparisons between a Pipistrel Sinus and the Merlin PSA

	Pipistrel Sinus (Kit)	Merlin PSA (Kit)
Payload	471lb	300lb
Construction	Composite	Metal
Stall Speed	35knots	33knots
Cruise Speed	111knots	90-120knots
V_{ne}	118knots	130knots
Rate of Climb	1063fpm	1400fpm
Ground Roll	460ft	150ft
Endurance	5hrs	4hrs
Wing Span	49ft	25.5ft
Height	6ft	6ft 11in
Length	21ft 4in	17ft 5in
Cockpit Width	44in 2-seats	27.5in 1-seat
Empty Weight	661lb	415lb
Gross Weight	1213lb	715lb
Fuel Capacity	15.9gal	13gal
Build Time	**400hrs**	**40hrs**

Table 1. Target Platform vs. Surrogate

Remember, that a factory-built aircraft, comparable to the Sinus kit is almost $200,000 and can't be modified without re-designating it as an EXPERIMENTAL – Exhibition category aircraft. This limits the mods that can be done to it and will require continuous approval by authorities every time it flies outside of the home base. The EXPERIMENTAL – Amateur Built category is far more flexible. Therefore, for us to make quick changes we'd have to build our own Pipistrel. That means a kit, either Pipistrel Sinus or something else. I'm not comfortable with a 400-person-hour Pipistrel Sinus kit compared to a 100-person-hour Merlin kit. The total cost for two Merlins, one with floats converted to fuel tanks and the other unmodified is less than the cost of a single Pipistrel Sinus motor glider.

Composites vs. Metal

The Pipistrel is made of composites, primarily Carbon Fiber while the Merlin is all metal except for the engine cowling and a few nonstructural components like wheel pants and fairings.

Composites have the benefit of being able to assume complex shapes and very smooth lines while metal parts are generally less "curvy."

Composites are the new materials and therefore there is a lot of pressure to make everything out of composites. The Pipistrel has its roots in the glider world which is almost exclusively composite owing to the strength and smooth surfaces achievable by this manufacturing method over the riveted aluminum aircraft common before composites. But composite construction has disadvantages in the prototyping domain when an existing design needs to be modified. That's because a composite structure can't be cut or drilled without compromising the integrity of the entire structure. For example, when we wanted to add something as simple as an antenna it required bonding 6 layers of fiberglass cloth to the location before putting in holes. We then had to bond copper to the outside surface to create a proper ground plane for that antenna. With an aluminum skin, you place a doubler over the hole to give the location some stiffness and they can be attached with the same fasteners as the fitting or antenna itself.

Durability:

A composite structure when damaged requires extensive work to repair. This includes sanding out the area, cutting out the defect, feathering the surrounding area to prepare it for the patch, laying a new patch which may require many layers of new fabric, and waiting for the bond to cure. With an aluminum structure a dent need not be repaired and patching an area usually means riveting on the patch of the same thickness as the damaged area.

Weight:

Despite claims to the contrary, in the homebuilt world where there are no autoclaves and money for high-grade carbon fiber, Aluminum structures weigh 20% less than composite structures of the same size.

Finishing:

Composite structures require extensive labor to finish and require special paint with UV tolerance and solar radiation capability. 50% gray is the darkest you can paint a composite. Aluminum has no restrictions on gloss or color.

Figure 13. Autopilot Tab for Merlin

Cost:

Aluminum structures start as flat panels and are formed to shape. With CNC laser punches these panels can be fabricated in minutes from computer data, data that can be changed in no time. Composites require molds which are expensive and almost impossible to change. A typical composite aircraft has about 100 molds. A typical aluminum aircraft has 6. Because of the overhead in tooling composite aircraft are more expensive than aluminum by as much as 10 times but more typically 3 times more expensive after most of the tooling has been amortized. Aluminum is a well-understood and easy to fabricate material. There is no special training to fabricate an aluminum structure. Composites require special skills.

Availability:

Aluminum aircraft kits can be fabricated in a matter of weeks and increasing production is much easier because there are no mold bottle necks. In the composite world the molds pace production. In aluminum construction increasing production means adding more hours or machines to the program. There are no special tools or molds to make.

Lead Time:

Like availability, the aluminum aircraft industry can easily adjust to changing market conditions thereby maintaining good lead times. Aluminum kits rarely exceed 3 months of lead time to have in the shop.

For a test program, the Merlin makes a good platform because it remains piloted even when it's being operated as a UAV. The Merlin uses an autopilot that uses PWM commands to operate the actuators. These are the same commands used by most UAV autopilots.

For the A3R® test, we will be using an open-source PWM autopilot which uses an open-source executive which is easily modified.

For A3R® we can connect the autopilot directly to the servos built into the Merlin bypassing the autopilot supplied with the airplane and with the RTK added be able to maintain the relative position of one aircraft to the other within 2cm over several hundred meters on the internal radios or miles with more powerful radios.

The autopilot on the Merlin is unique because it uses tabs on the control surfaces rather than driving the control surfaces, directly. This means that there are no complex linkages, emergency release clutches, or expensive servos to deal with. The tabs are flat metal plates connected to small servos like what you see in small UAVs. To the pilot, the forces feel like trim changes which are, in fact, how they work. The other nice thing about this arrangement is that it easily supports the use of various autopilots simply by connecting them to the control wiring for a servo. We can give the pilot the option of using the onboard autopilot, UAV control, or none with a toggle switch. In any case, the pilot will always be able to assume full control of the aircraft.

Figure 14. Merlin Amphibian Tanker with Merlin PSA Receiver

I think a Merlin with a pilot onboard is an unbeatable test platform for almost everything we want to do with a UAV. We can test controls, radios, payloads, CONOPS, external stores, software, and even new wings, etc. They have a BRS so there is a huge measure of safety.

So, if we go with a Merlin how do we simulate an actual A3R® system? The trailer testing has already given us the locations of the two main aircraft.

The tanker will be the Merlin Amphibian and aircraft that's already flown and with a built-in platform for stowing and managing a dual end effector system. This is what a Merlin-to-Merlin A3R® system will look like.

Figure 15. Initial Position

The tanker takes its commands by flying to a position relative to the receiver. It isn't controlled by the receiver but corrects its flight path to zero out deviations from the "ideal" position as reported by the navigation system. The cameras in the back of the pontoons on the tanker can also provide location data for a backup. Once in a stable position ahead of and in front of the receiver, the tanker deploys the end effector(s).

Figure 16. End Effectors Disconnect from Autonomous Tanker

Here you see them as they disconnect from the tanker. The process is simple, the end effector opens its roller clamps and it's free. The hose reels inside the tanker pay out the hose as required. They don't leave their docking station unless they have visual contact with the receiver.

Here you see them tracking and flying to their respective fueling blades on the receiver. The pontoons aren't as far apart as those locations but there is plenty of control authority to move laterally to their targets which were confirmed by the truck tests. Note, the speed brakes are out on the little wings. The speed brakes are the throttle

Figure 17. End Effectors Perched and Dispensing Fuel

control for them. Testing was to put the two heavy players in a position in the middle of the control authority of the speed brakes, not the middle of their travel. The 4 cameras on the end effector have acquired all the fiducial markers on the top of the receiver's wings. Fiducials aren't mandatory but they aid in navigation and identification of the receiver.

The end effectors are directly over the fueling blades and won't commit to a contact until both are in a stable position. Because the end effectors are very light, they are also very maneuverable. They have no trouble holding position relative to the receiver. Here is what they look like just before they clamp.

The roller clamps are not moving and won't move until both are locked onto the refueling blade. When they do clamp it is quick and precise.

The servo motor drives the rubber-coated rollers through a planetary gearbox with plenty of torque to drive the hose down into the well. It also seals the vent to the vent line. The valve plunger is long and makes good ground before fuel is allowed to flow to dissipate static charge.

Once the A3R® is seated on the fueling blade the flow begins immediately because there is no mechanical or electrical valve to begin flow. It's all done by the downward motion of the end effector onto the fueling blade. The emergency disconnect is for the rollers to open at which point the spring-loaded valve pushes the end effector up and off, instantly. This safety feature is crucial and available with a non-locking, yet positive sealing mechanism. Neither the probe and drogue or boom and receptacle methods can use the engaging motion to provide a valve opening, sealing, release and seal all by the motion of the clamp.

By using two Merlins to do the system validation testing, we can introduce an entirely new way of air-refueling to the warfighter. But we can do more.

CONCLUSIONS

Using the amphibious version of the Merlin is a good simulation of a ramp-based docking station which allows us to easily scale the system up to any commercially available amphibian but also to something as large as a C-27, C-130, C-17, or C-5 with a ramp. The pontoons can demonstrate docking, launching, hose management, contact, recovery, and stowage as if it were a cargo ramp. Even the Cessna 208B Caravan is often equipped with amphibious pontoons.

There is provision for electrical power to the docking station to recharge and hold the end-effector in place, in an aircraft with an open cargo hold and ramp we can scale the A3R® up or provide a swarm coming off the back for a variety of tethered operations.

A transport class aircraft can use the A3R® techniques to launch and recover test instruments, weapons, rescue grapples, and even massive fuel offloads not even available with current methods. The receivers need only have a fueling blade or compatibly shaped fixture to become instantly air-refuellable or rescuable. The receiver doesn't need to be trained to intercept, fly to, or especially initiate the contact. That's up to the A3R® system which needs only a marginally cooperative target to work.

A stump on the top of aircraft will have minimum drag and can be placed anywhere on top of an aircraft that can be reached by the end effector. The airspeed band of a A3R® is far greater than that of any other technology. There is no limit to what speed it flies if the location is mapped and accounted for to make a contact.

ACKNOWLEDGEMENTS

I would like to extend my sincere appreciation to Mr. David Solomon, Director of Technology Transfer for providing guidance and data for this effort, Mr. Russell Wolfe, VP-Engineering for allowing me the time to develop this report and to my late father, Marinus A. Bosma for his inspiration and support as an inventor, warrior, and mentor.

REFERENCES

1. Tanner, R. M. (2006), *History of Air-to-Air Refueling*, Great Britain, Pen & Sword Aviation
2. Bosma, M. B. (2020), *Refueling System and Method*, USA, US Patent 10,654,584
3. Bosma, M. B. (2022), *Refueling System and Systems with End Effectors*, USA, US Patent 11,505,330
4. Bosma, M. B. (2023), *Helicopter Cliffside Rescue*, USA, US Patent 11,639,233
5. Bosma, M. B. (2023), *Helicopter Winch Cable Stabilizer*, USA, US Patent 11,618,584
6. Bosma, M. B. (2023), *Satellite Refueling*, USA, US Patent 11,623,760
7. Bosma, M. B. (2023), *High Bandwidth Data Transfer*, USA, US Patent 11,634,237
8. Bosma, M. B. (2023), *Vehicle Refueling*, USA, US Patent 11,591,106
9. Bosma, M. B. (2023), *Fire Bomber Delivery*, USA, US Patent 11,591,105

BIOGRAPHY

Marinus B. (Ben) Bosma graduated from the United States Air Force Academy in 1975 with a degree in Engineering Sciences majoring in Space Vehicle Design with minors in Astronautical, Aeronautical, and Mechanical Engineering. After graduation, he attended undergraduate pilot training at Vance AFB, OK. His first assignment was as an instructor pilot. He then went on to fly the F4D, F4E, and F4E-ARN101 Phantom in the USA and Europe. He transferred to the Air National Guard where he completed his time in the A7D Corsair II. While in the ANG he was the Vice President of Engineering at Bosma Machine and Tool Corporation where he received his first patent in a manufacturing system for precision radio telescope reflectors for the astrophysical community where he was a member of the Event Horizon Telescope team. He has since been issued over two dozen patents in manufacturing, astrophysics, medical equipment, sports, aircraft, hybrid propulsion, rescue systems, and tethered systems for aircraft, ships, ground, and space applications. He was the lead architect of several autonomous aircraft programs for the Air Force Research Laboratory. He's an avid homebuilder having built two aircraft and is working on his third. He flies regularly in his Cirrus SR-22TN and has been the Learjet pilot for chase and test pilot school support at Edwards AFB, CA. He has flown over 60 aircraft and maintains a Commercial, Multi-engine, Sea, Land, and Instrument Rating. In addition to flying Ben is an avid SCUBA diver. He resides in Tipp City, OH with his wife of 47 years, Ginny.

Remote Measurements of Ocean-Going Vessel Fuel Sulfur Content

Acknowledgments

This project funded by the US Environmental Protection Agency (EPA), U.S. Department of Transportation Maritime Administration (MARAD), and California Air Resources Board (CARB). It was conceived, organized, and coordinated by the Adept Group, Inc. ADEPT). Project participants include CARB, UCR, CE-CERT, UCLA, Cal Maritime Academy, AEROMON/NUaer, Flying Lion Inc., TELOPS, Hitachi, LA Bunker Surveyors, Inc., Maersk, EPA, MARAD, and USCG.

Introduction

The project builds on prior extensive work conducted in Europe (Belgium, Denmark, Finland, Norway, and Sweden) to monitor the sulfur content in the fuel burned by Ocean-Going Vessels (OGV's) as well as on prior work at the University of California Riverside (UCR) to test and evaluate means to verify air quality regulatory practices in the maritime sector.

The regulatory environment affecting the United States (US) maritime industry is increasingly restrictive – including air quality regulations. The European Union (EU) authorities and their private sector partners have led in this field – where large budgets were allocated over the last ten years to develop and test aerial monitoring of OGV emissions. Thrse multi-nation applied R&D efforts resulted in the deployment of dynamic measurement systems in several EU air quality sensitive areas - yielding positive results. In contrast – in the US – until now, relatively little was spent on similar research. As a result – despite having signed a binding agreement [1] with the International Maritime Organization (IMO), an arm of the United Nations (UN) – the US lags in the deployment of such systems. Part of this lag is attributed to the need to customize such systems to US specific geographic, regulatory, traffic, and economic conditions.

This was a verification as well as an enforcement focused project. Its results were meant to inform policy makers at international, federal, state, and regional level – and to facilitate when appropriate – the adoption of one or more novel methods for more comprehensive monitoring of FSC regulations in Sulfur Emission Control Areas (SECAs) – including California waters (where the California OGV Fuel Regulation applies). Thanks to its unique construct – the project yielded highly valuable data for the maritime industry across the world as well as for those charged to construct and/or monitor existing and anticipated air quality regulation for this industry – as well as to help ensure there's a level playing field for those who do observe such regulations.

Four systems to measure fuel sulfur content (FSC) were tested – two of which are remote sensing systems. Essential data was collected in August 2022 on a container vessel (Gerner Maersk) operating in California waters under controlled conditions. Measurements were taken over a full day while the Gerner Maersk first ran on low sulfur marine distillate fuel (> 0.1% S), then on higher sulfur fuel (0.5% S) – and finally back to low sulfur marine distillate fuel. The three on board FSC measurement systems were:

(i) Unmanned aerial vehicles (UAVs) equipped with a customized gas sensor box,
(ii) Portable hyperspectral camera, and
(iii) Direct stack raw gas sampling and quantification.

These systems were operated at the same time.
The respective FSC results were compared to the standard method (bench X-ray fluorescence (XRF) method, ISO 8754) now practiced by CARB. A fourth measurement system (e.g., Hitachi portable XRF instrument) was tested a week later on a representative sample of the collected fuel samples.

Further, CARB requires OGV's to use distillate fuels (> 0.1% S) such as a marine gas oil (MGO) within 24 nautical miles of California coastline. At CARB's request – additional work was conducted to verify the "distillate" requirement in its regulation.

Executive Summary

This unique and never before attempted project's underlying idea was to simultaneously collect data from all the methods under test – and compare the results vs. the actual sulfur content in the fuel. The sulfur content of the fuel was derived via extracting fuel samples every 15 minutes from a point as close as possible to the OGV's main engine (ME) while the OGV is under way – and having these same samples analyzed by two highly respected laboratories using a method approved across the world (ISO 8754).

Although based on a simple simultaneous testing concept – this testing method's comparative exercise turned out to be extremely complex to organize and carry out. Numerous hurdles were overcome - from test vessel switches to major port congestion delays, COVID-19 barriers, to unexpected equipment failures, to difficult flying conditions around and in the air space of a large container vessel under steam – just to mention a few.

The primary objective was to verify several novel technologies/tools [e.g. (i) UAV emissions detection and quantification device - which was previously tested/used in the EU; (ii) a Canadian high end hyperspectral camera which has shown promise in the St. Lawrence Seaway; and (iii) in-situ emissions monitoring) versus established protocol in-lab measurements using X-Ray Fluorescence (XRF).

This work has shown that one such tool is ready for adoption (e.g., portable XRF instruments). The collected data also supports that remote sensing systems mounted on UAV platforms are worthy of follow on work to optimize their field use. The other remote sensing system (e.g., hyperspectral camera) installed on the test vessel encountered unexpected interference in its data collection efforts. This experience falls under "lessons learned" and should not discourage further exploration of this method given its performance in other circumstances.

Valuable data was also collected on the time needed to fully switch to meet the distillate fuel requirement for steaming in California waters. This data indicated that it takes longer than anticipated to date to achieve such status.

Based on the collected results a follow-on project, also to be conducted in California waters but over a longer period, is indicated. Unlike this project – the remote sensing systems are to be based on land near the shipping lanes or on a strategically positioned chase vessel.

Objective

This research was undertaken to assess the feasibility of several low-cost remote sensing processes to identify non-compliant OGV's for subsequent on-ship CARB fuel inspection. As part of this work, a team of scientists, engineers, and technicians explored the use of UAV mounted sensor systems to quantify the sulfur content in the fuel burned by an OGV – be it at dock or at sea. The directly measured FSC's or the calculated FSC's were compared to the FSC data from the XRF benchtop instruments used by CARB and by Los Angeles Bunker Surveyors (LABS) laboratories.

Materials and Methods

Three sets of emission measurements were concomitantly taken with three instrumented systems aboard the Gerner Maersk. Measurements were taken while the vessel sailed mostly in a straight line at one speed (~15 knots at 22–23% engine loads) while the Gerner Maersk's ME first ran on low sulfur fuel (LSF), and then on higher sulfur fuel (HSF) – and lastly again on LSF. The same emissions measurement protocols were followed while the OGV ran on LSF and on HSF. The data collection took place over 11 hours and 36 minutes with the on-board instrumentation setup one day earlier.

Emissions were measured with CE-CERT's instruments using two probes inserted in the plenum below the OGV's economizer. Gaseous emissions (including SO_2) were measured from the raw exhaust using a Horiba PG-250 portable multi-gas analyzer. The Horiba simultaneously measures the separate gaseous components (CO, CO_2, NO_x, and SO_2) using ISO/IMO and USEPA recommended methods. The output signal is then connected to a laptop using an RS-232C interface that continuously records data measurements. Analyzer checks were performed using calibration gases prior to and after each test to check for drift.

Simultaneously, sensors mounted on a UAV detected and measured several gases at three predetermined distances from the stack. The measurements collected from the gas sensors on the UAV were used to calculate the FSC in the fuel burned by the Gerner Maersk.

A TELOPS hyperspectral camera was also mounted on the top deck of the Gerner Maersk. It was meant to measure FSC while all the other measurements were going on. Regrettably – the hyperspectral did not yield usable results.

While this was going on - fuel samples were collected next at the ME every 15 minutes. These fuel samples were tested by CARB's laboratory as well as by a private certified laboratory in accordance with ISO 8754. FSC's were also collected from analyzing some of the same fuel samples with a Hitachi portable XRF instrument.

Several tasks were conducted by the technical team over the course of the project:
Task 1: Preparatory work including vessel site visits to assess logistic requirements for sample collection and drone deployment, writing Sensor Package operator training manual and preliminary remote on-land UAV pilot training (surrogate measurement environment), remote learning and coordination with the sensor box (a.k.a. BH-12) operator.
Task 2: Identify crew/vessel requirements that affect integration with existing instrumentation to ensure synchronized data collection and recording. Set up data recording systems that interact with the OGV's operating systems, considering at sea conditions- as well as sensor cloud communication, storage, and data retrieval with data backup system.
Task 3: Prepare and submit a comprehensive Work Plan to the Gerner Maersk's Captain and to Maersk's execs - as well as to CARB, US EPA, MARAD and USCG representatives.
Task 4: Conduct virtual training for the sensor system operation, and then test sensor system under field conditions to maintain accurate positioning of the sensor package within the exhaust plume during measurement, to determine accurate feedback communication for the UAV pilot during FSC sea trial measurements.
Task 5: Maintain dialogs with EU scientists leading similar projects and US agencies involved with FSC compliance enforcement to:

a. Leverage their experience & know-how to avoid or overcome potential challenges and help analyze the collected data.
b. Gain valuable knowledge on how similar detection and monitoring tools under development are applied in the field.
c. Gain additional knowledge of desired features and specific requirements that must be met for US agencies to adopt use of the technology platform.
d. Understand enforcement protocol concerns.

Results

The FSC data generated by all the performing methods is shown in Figure 1. (Note: The TELOPS hyperspectral camera did not produce any usable data).

The fuel samples tested by the CARB lab and the private lab per ISO 8754 method (Figure 2) were closely matched and corroborated Gerner Maersk's bunker reports. These two data sets had an average difference of 0.051 wt % with a range from -0.081 to 0.024. The private lab tested a representative sample of fuel samples – and the ensuing results closely matched the CARB lab FSC results.

Fig. 3 compares the CARB's ISO 8754 results vs. Hitachi XRF instrument results. The Hitachi instrument results had the smallest average deviation from CARB's results of 0.0080 wt% and a difference range from -0.0025 to 0.0239.

The UAV sensors AEROMON method is compared directly to CARB's lab results in Figure 4. The deviation ranged from -0.080 and +0.071 with an average difference of 0.046. UCR's ISO 8178 method had the largest deviation average of 0.088 and widest range of -0.183 and +0.203

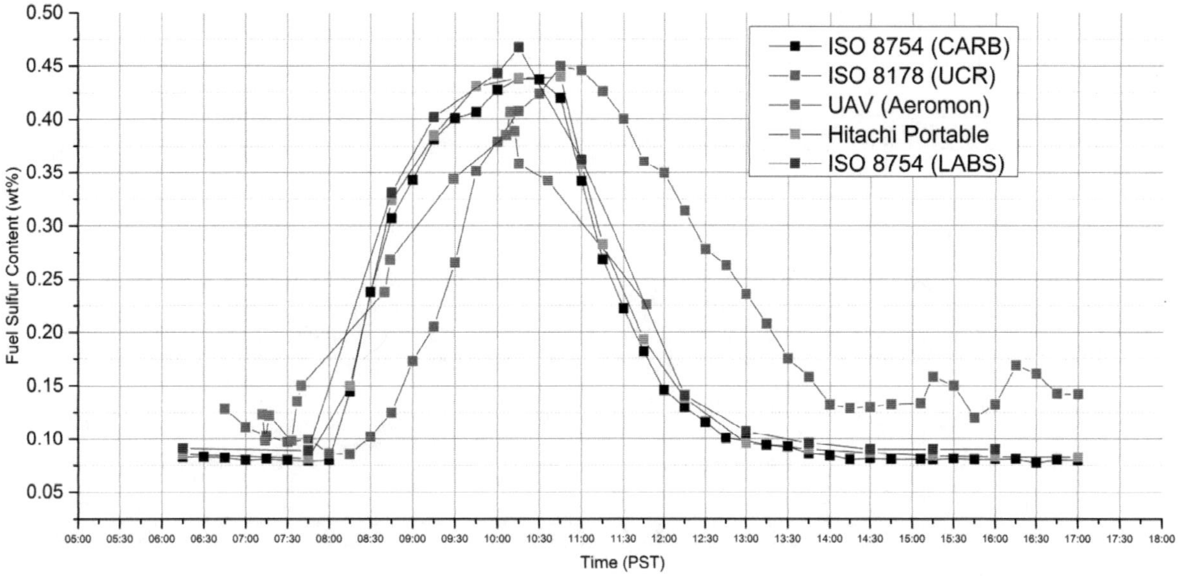

Figure 1. FSC (wt%) vs. time for all sulfur content measuring methods- as tested on the Gerner Maersk.

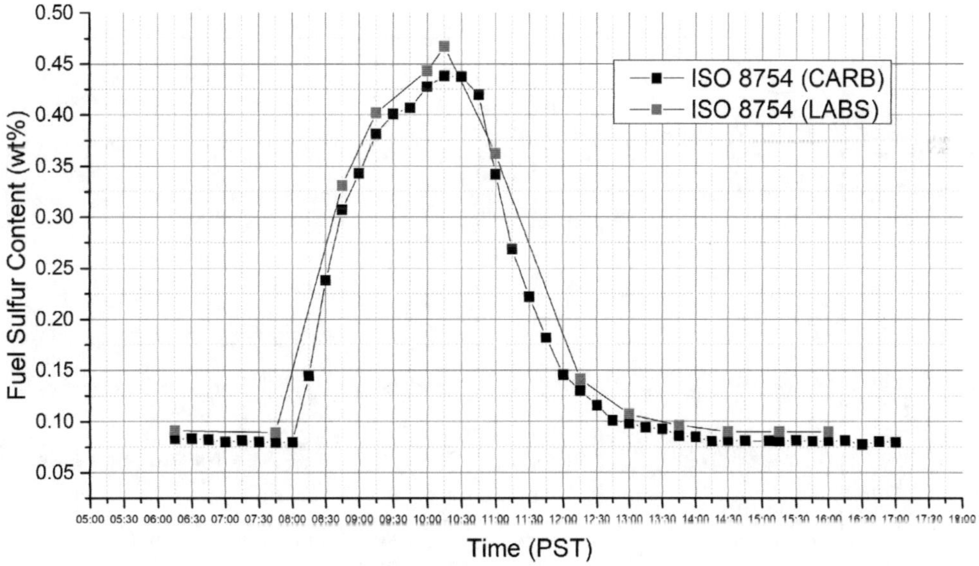

Figure 2. Direct comparison of CARB's ISO 8754 method vs. LABS's ISO 8754 method.

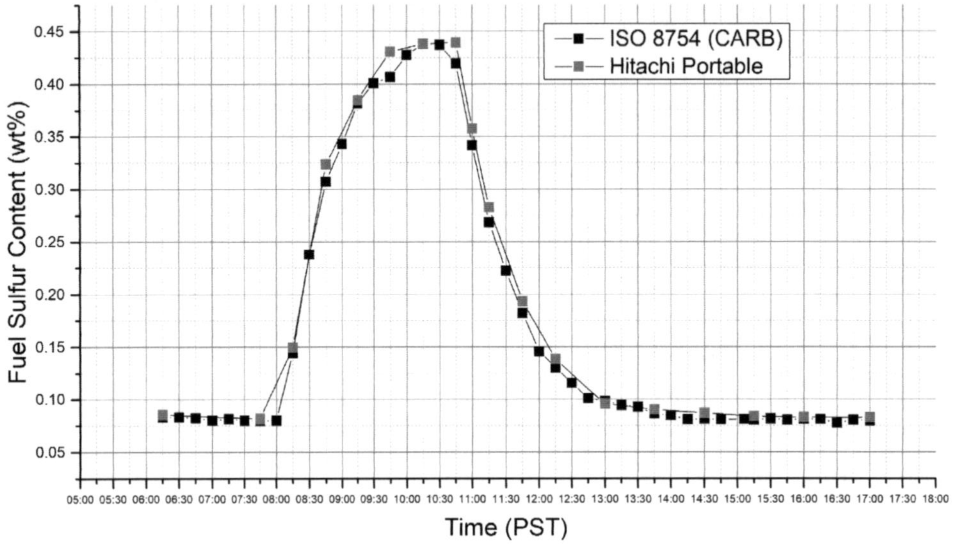

Figure 3. Direct comparison of CARB's ISO 8754 method vs. Hitachi Portable XRF instrument.

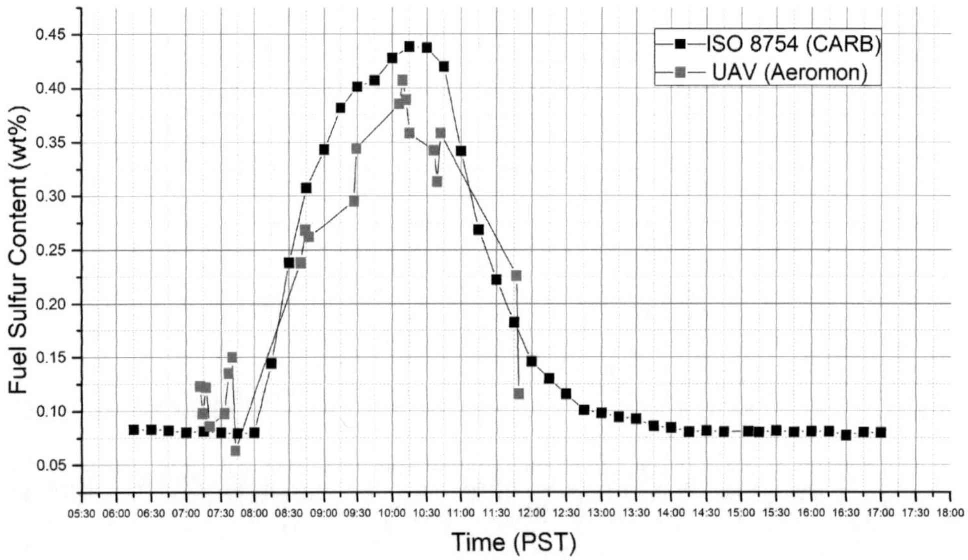

Figure 4. Direct comparison of CARB's ISO 8754 method and Aeromon's UAV sensing method.

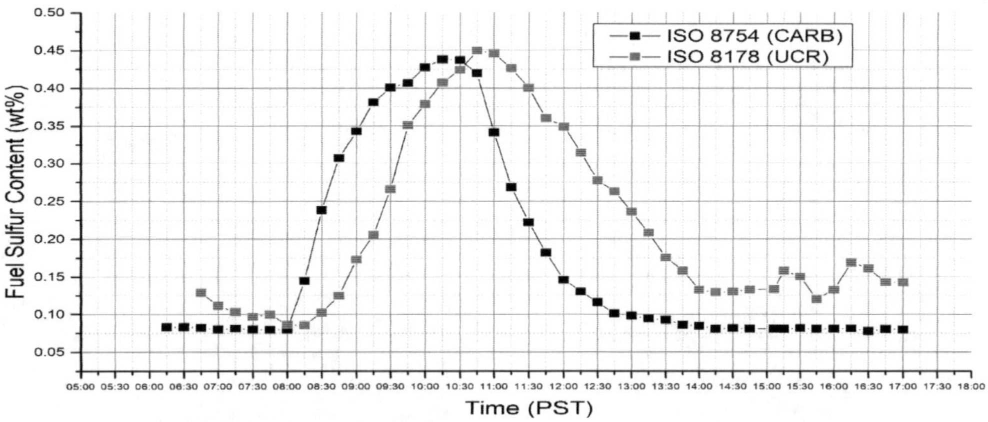

Figure 5. Direct comparison of CARB's ISO 8754 method and UCR's ISO 8178 method.

Fuel Compliance Discussion

The CARB OGV Fuel Regulation requires the use of marine distillate grade fuel (marine gas oil or marine diesel oil) with a maximum sulfur level of 0.1% while operating auxiliary diesel and diesel-electric engines, main propulsion diesel engines, and auxiliary boilers on OGV's within all California waters within 24 nautical miles of the baseline - RCW). Ships operating within the California RCW must use distillate fuel as per ISO 8217 (Title 13 CCR sect. 2299.2 subsect. (i)(1)).

Per Annex VI of the IMO International Convention for the Prevention of Pollution from Ships (MARPOL), the limit for sulfur in fuel oil used by OGV's operating outside designated emission control areas is 0.5% sulfur, and within an ECA, it's 0.1% sulfur. While the fuel sulfur limits under the CARB OGV Fuel Regulation and the ECA Regulation are the same (0.1% sulfur within 24 nm of shore), the CARB regulation requires distillate grade marine fuels, while international standards allow ultra-low sulfur fuel oil, which is not a distillate fuel. The removal of sulfur from fuels prevents the formation of sulfur oxides in engine exhaust, which in the atmosphere becomes ambient sulfate particulate. However, the use of distillate fuel also reduces the formation of directly emitted particulate from diesel engines, as the fuel burns cleaner.

Visual observation is one simple method that the fuel being sampled may not meet the distillate standard. Uncontaminated distillate fuel has a *"clear and bright"* appearance, as defined by ISO 8217. Distillate grade fuels contaminated by residual grade fuels have an opaquer appearance making any level of contamination easily identifiable. See Figures 6 – 9.

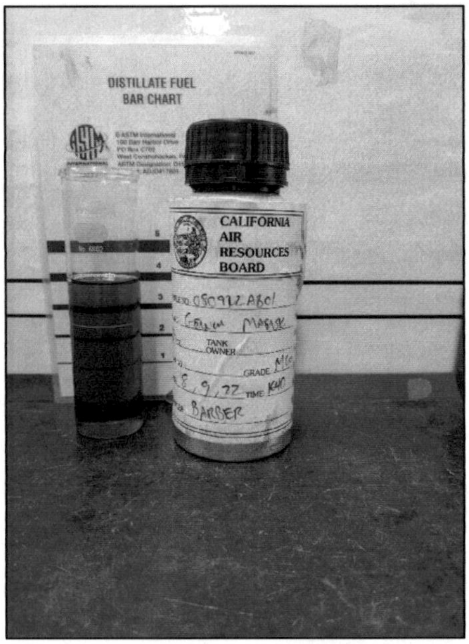

Figure 6. 080922AB01 represents the clean sample of uncontaminated MGO at the beginning of the test day which passes D1476.

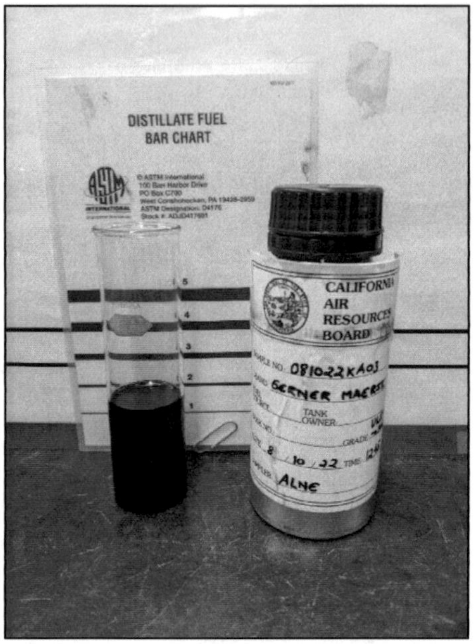

Figure 7. 081022KA03 taken at 1245 hours, represents the sample of MGO just approaching the time of sulfur compliance.

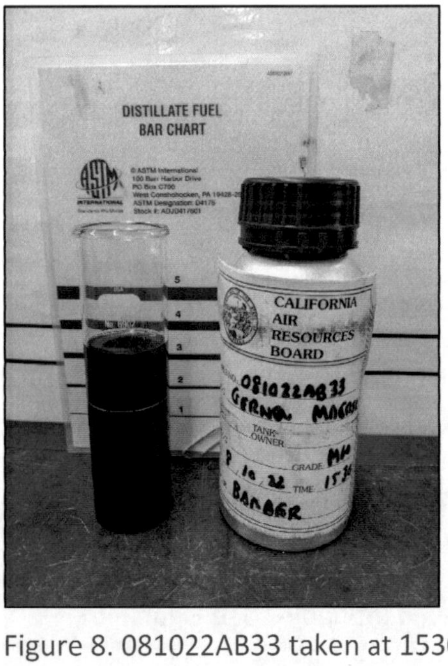

Figure 8. 081022AB33 taken at 1530 hours, represents the sample of MGO just approaching the time of Carbon Residue compliance.

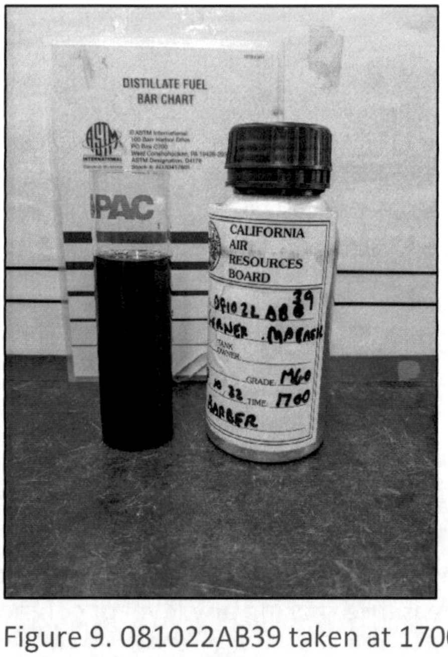

Figure 9. 081022AB39 taken at 1700 hours, is the final sample of the day, still showing considerable opaque appearance that is off the scale for D1476.

The ISO 8217 also specifies ISO 10370 for Carbon Residue testing; Micro Method. In the test protocol, a weighed quantity of sample is distilled to the bottom 10% and placed in a glass vial and heated to 500 °C under an inert (nitrogen) atmosphere in a controlled manner for a specific time. The sample undergoes coking reactions, and volatiles formed are swept away by the nitrogen. The carbonaceous-type residue remaining is reported as a percent of the original sample as "carbon residue (micro)." The limit for carbon residue in ISO 8217 is 0.30%.

This project provided a unique opportunity to explore fuel contamination, and the methods used to calculate times required for compliant operation within California waters. Fuel changeover calculators, the most common being the one provided by the Fuel Oil Bunker Analysis Service, or FOBAS, are used all over the world to determine the time required to switch from fuel of one sulfur level to another to meet the compliance goals of an ECA. Where the limits include a distillate marine fuel requirement, the FOBAS may not be the most effective tool to determine the fuel changeover calculation times. This project's results indicate that more time is required to comply with the distillate component of the California's regulation.

In this study, sulfur compliance was met at approximately 13:00 hours, while MCRT compliance was achieved at about 15:45 hours. Fuel consumption rates varied considerably during the project; onboard ship calculation estimated a fuel change over time of approx..3 hours to meet sulfur compliance. This calculation onboard the Gerner Maersk achieved the goal for sulfur compliance in that time frame but did not meet the specifications of distillate grade as outlined by the CARB OGV Fuel Regulation. See Figure 10.

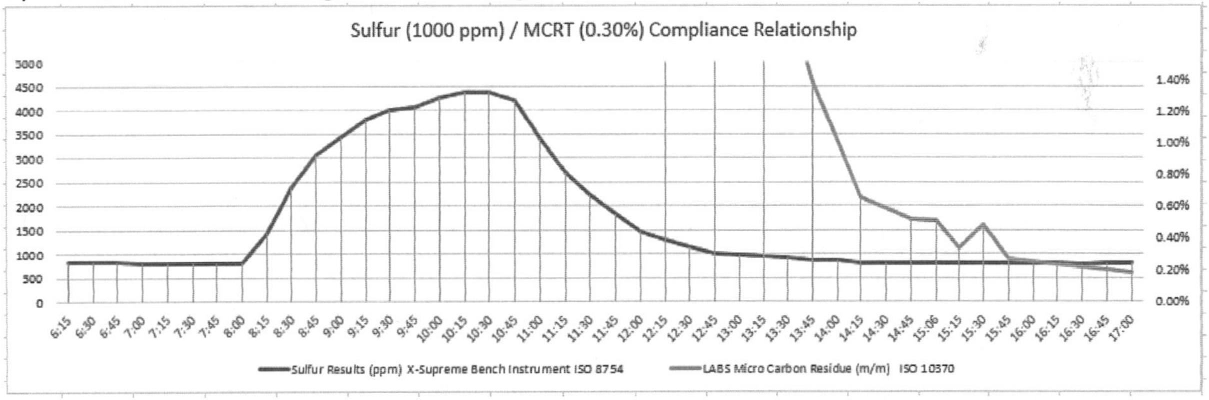

Figure 10. Sulfur test results (left axis, ppm) and carbon residue test results (right axis, wt%) versus time from sampling at the final duplex filter on the Gerner Maersk.

Discussion

The project hypothesized that one or more novel approaches can cost-effectively, timelier, and more comprehensibly measure FSC while OGV's are under way or while at pier. First - the collected data clearly indicates that the Hitachi portable instrument provides reliable results that are also timelier than the results provided by CARB's lab. Adding this instrument to the tool set of the CARB (or US Coast Guard) inspectors would save professional time and allow for more fuel sampling and timelier corrective action – when such action may be needed. There is

the added benefit of being able to have the vessel representative – if appropriate – view in real time of the fuel sample testing process.
Second – the data collected via the AEROMON UAV + sensor box method has shown that it holds the promise to be a time saving and viable method to assess FSC compliance to a greater degree than the currently practiced method. At the same time – in its current state of development – reliable results are predicated on the UAV being flown by a competent pilot within a given range from the vessel's stack.
Prior to adopting AEROMON's UAV system to enhance California's Sulfur Rule – as to address monitoring of OGV traveling in **and out** of California's ports (as of now – it is impossible to assess such compliance once the vessel has left the pier) – follow-on work is indicated over a representative period in high traffic waters (e.g., Santa Barbara Channel).
Third – given the uncovered disparity of results from the ISO 8754 protocol and UCR's ISO 8178 protocol – additional work is indicated to address the concerns behind it.
Lastly – despite the lack of usable results from the hyperspectral camera experiment – additional work with this technology is indicated given prior promising results in the St. Lawrence Seaway

Summary and Conclusions

The Hitachi portable XRF instrument proved to be the most accurate of the new methods tested. Its adoption by CARB, USCG, and EPA inspectors is recommended.

Follow-on work with the Aeromon box on a longer flight duration UAV is recommended over high OGV traffic waters.

Land based near-shore OGV testing with the TELOPS hyperspectral camera is also recommended.

Lastly, the discrepancy in the CE-CERT FSC results versus the XRF FSC results merits further investigation.

References

[1] US Coast Guard /US Department of Homeland Security CG 543. Policy Letter dated February 9, 2009 https://www.epa.gov/sites/production/files/2016-09/documents/cgpol0901.pdf.

Appendix A – Sampling Methodology

CARB Fuel Sampling Procedure:

(see MEPC.1/Circ.864, International Maritime Organization "Guidelines for onboard sampling for the verification of the sulfur content of the fuel oil used on board ships")

The on-board representative sample or samples should be obtained from a "designated sampling point(s)", the location identified by the vessel management company as representative, or another location agreed on by the CARB inspector and the ships representative on the vessel being inspected. It must be noted that at times "designated sampling points" are not representative and should in some cases be avoided.

Representative sampling point(s) to be used should fulfil all of the following conditions:

1. Be easily and safely accessible.
2. Be downstream of the in-use fuel oil / diesel oil service tank and downstream of the three-way fuel oil / diesel oil valve.
3. Be within the system of recirculation between the combustion machinery and the fuel oil supply module, as long as they are downstream of any three-way fuel valve, within the system of recirculation, and easily flushable.
4. Be as close to the fuel oil combustion machinery as safely feasible taking into account the type of fuel oil, flowrate, temperature, and pressure behind the selected sampling point.
5. Be in a position shielded from any heated surface or electrical equipment and the shielding device or construction should be sturdy enough to endure leaks, splashes or spray under design pressure of the fuel oil supply line so as to preclude impingement of fuel oil onto such surface or equipment.
6. Be accepted by both the ship's representative and the inspector.

Fuel oil samples may be taken at more than one location in the fuel oil service system to determine whether there is a possible fuel cross-contamination in the absence of fully segregated fuel service systems, or in case of multiple service tank arrangements.

Examples of good sampling locations include but are not limited to Final Filters for Main Engine and/or Generator Engine Fuel Oil Supply Module(s) Final Filters, fuel line valves designated for sampling just before the combustion machinery, and Final Filters on combustion machinery.

In situations where the fuel being sampled has an opaque, dark, and / or waxy appearance, additional sampling may be required. On the Gerner Maersk, sampling was conducted from the Main Engine Final Duplex filter in operation at the time of this exercise.

Aeromon/FlyingLion UAV Methodology:

Aeromon BH-12 mini sniffer

The Aeromon BH-12 emission measuring device (Figure 11) is a modular and portable piece of equipment intended for simultaneous detecting, measuring, and mapping of multiple airborne gaseous compounds, particulate matter, and noise. The device communicates its location and measured values in real-time to Aeromon Cloud Service (ACS) which is used to store, analyze, and visualize the results. BH-12 supports various datalink options and is normally used either with IP radio or direct mobile LTE connection.

Figure 11. Aeromon BH-12 mini sniffer

The sensor technologies utilized for this measurement campaign were Non-Dispersive Infrared (NDIR) for CO_2 and Electro-Chemical (EC) for SO_2, NO and NO_2.

Calibration procedure

Aeromon performs factory calibration with certified gases when sensor modules are assembled. After this the sensor modules are stored in controlled conditions. When sensor modules are used in field ops the calibration level is checked daily (calibration before starting and level check after measurement) with certified calibration gases and known cross interferences (including humidity) to maintain result traceability. This field calibration or calibration level check only takes a few minutes to complete, and the procedure is guided step-by-step for users by BH-12 software. Results of the calibration and corresponding calibration factors are stored in Aeromon Cloud Service for traceability. When sensor response declines under pre-determined threshold levels the sensor modules are changed. Also, as sensor response to target gases and their cross interferences changes with T, RH and sensor ageing, this daily quality check procedure provides an essential quality control for small sensor, sample line and full system performance. The calibration gas concentration tolerances are included as a part of the uncertainty calculation.

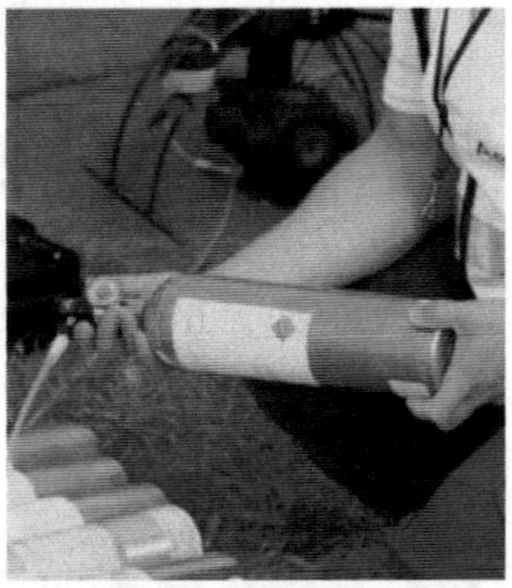

Figure 12. Daily calibration level check is always performed through the sample line in use.

Error sources accounted for reporting

Finnish Meteorological Institute reference laboratory (accredited according to ISO 17025) has performed an analysis of the expanded uncertainty of Aeromon system according to JCGM 100:2008 and the device performance characteristics have been defined following the draft technical document of CEN TC 264/WG 42: Air Quality – Performance evaluation of sensors for the determination of concentrations of gaseous pollutants and particulate matter in ambient air.

Aeromon has a minimum time window definition based on SO_2 sensor response time (slower than CO_2, NO, NO_2). When minimum time window requirement is exceeded, we calculate the FSC point-by-point within the FSC integration time window. The deviation of the point-by-point FSC result is a part of the uncertainty calculation. This analysis step takes into consideration both the signal-noise (S/N) ratio of FSC sensors and pilot performance during sampling time window. The combined uncertainty is calculated as follows:

$$\boldsymbol{U_{FSC}}(\%) = \sqrt{(\boldsymbol{U_{cal}})^2 + (\boldsymbol{U_{int}})^2 + (\boldsymbol{U_{FSCrel}})^2} \qquad \text{Eq. 1}$$

U_{cal} = calibration gas tolerance

U_{int} = intrinsic errors incl. sensor and device uncertainties from component manufacturer declarations and laboratory testing

U_{FSCrel} = relative error of FSC during the plume visit. The sampling or drone pilot performance and sensor S/N ratio are the main sources for this error. In cases where sampling concentrations vary notably during plume visit this error increases as different sensors react to altering concentrations at different characteristic speed, thus, giving rise to deviation in point-by-point calculated FSC value. In plumes with very low SO_2 concentration in relation to CO_2 the S/N ratio of SO_2 signal may become the dominant source of significant increase of this error.

The ambient baseline of all the measured gas components is continuously measured around the vessel plume and the background is removed from the result analysis. In general, the ambient level changes of CO_2, NO_x and SO_2 are typically very small in comparison to our typical sample concentrations (CO_2 > 150 ppm above atmospheric background, SO_2 and NO_2 in range of > 0.1 ppm and NO > 0.5 ppm above ambient levels. However, in the FSC analysis tool in ACS the user can adjust the background level (level is selected for both before and after the plume sampling) for each gas parameter individually if needed.

Data reporting

All the measurement data is visualized in real-time for the drone pilot and payload operator. Typical individual result reporting includes the result and its uncertainty paired with measurement time (UTC), location (lat., lon., alt.), device and sensor module identification and measurement target identification information, such as IMO number, MMSI number, vessel name, location, and other standard AIS data fields.

Plume approach

The ship plume is directed towards the vector sum of true wind and created wind, as shown in Figure 13. Plume altitude depends heavily on the stack height and geometry in relation to (apparent) wind speed and direction. After the plume direction and altitude have been identified the indicators for plume entrance are optical observation of onboard cameras (e.g., visual, thermal), drone experiences turbulence and Aeromon Plume UI indicates visually to pilot crew that plume entrance has happened. The Plume UI uses the threshold concentration levels of plume compounds and also indicates the pilot crew on real-time concentration level when minimum time needed inside the plume has passed. The aim of sampling is not to fly through the plume, but to find a position with high enough concentration and stay there for sufficient time, which depends on the slowest sensor T90 response time equipped onboard BH-12.

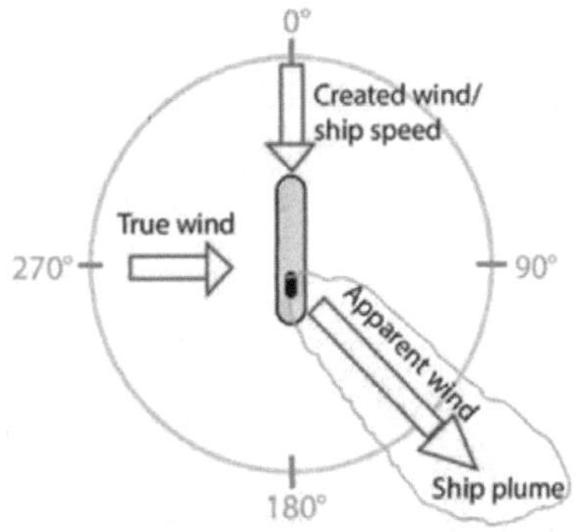

Figure 13. Ship plume direction (Berg et. al. 2012)

Campaign details

In the campaign the drone was piloted directly from the target vessel deck and the measurement distance was varied between 50, 100 and 150 ft distances in relation to exhaust stack. The lack of FSC results after the second fuel switch was because drone pilots required some rest.

UAV + Sensor System Trials

Four training and testing exercises were undertaken over three plus years to prepare for a one-of-a-kind mission – to wit: using a multi-gases sensor box mounted on a UAV - detect, optimally position, and measure gaseous emissions form the stack of an ocean-going vessel (OGV) that is under way - with the UAV launched from the OGV's deck. Fortuitously, after three unsuccessful attempts to conduct this work with different MARAD vessels, access to the Gerner Maersk was

kindly granted. These training and testing exercises were conducted to allow for the highest probability of success once at sea.
These exercises fell into three categories:

(i) Static dockside,
(ii) Replication of at sea OGV motion – desert work, and
(iii) In port, at anchorage, or transit within the port.

Despite extensive preparations, several challenges were discovered once under-way onboard the Gerner Maersk

Static Dockside

In May 2019, tests were run with a sensor box supplied by Explicit, a Danish firm, while TS Golden Bear was docked at Berth 46 in the Port of LA. For this test, the TS Golden Bear ran all its auxiliary generators to create as large as possible plume.

The 2019 tests uncovered significant challenges to locate the plume and the higher emission concentrations within the plume (a.k.a. the "sweet spot"). To overcome this, an infrared (IR) camera (sensor) was added to the UAV platform. This IR sensor can allow for optimal plume source and plume shape detection.

Subsequently – there were multiple dockside tests with MARAD vessels in Oakland and Long Beach. They included – and were not limited to:

- August 29, 2020 – IR sensor tests to locate detect plume and "sweet spot".
- July 19, 2021 – Test range finder's ability to determine distance from stack.

Desert (Constant Velocity) Tests to replicate at sea OGV conditions.

In 2021- tests were conducted on the dry lake at El Mirage on:

(i) August 14,
(ii) August 31
(iii) October 6, and
(iv) November 18

The systems repeatedly tested included - and were not limited to:

(i) Flight Operations Protocol.
(ii) Communication Protocol; and
(iii) Aircraft performance

Several challenges were identified and resolved:

Four UAV models were flown and evaluated for compatibility:

- DJI M600
- DJI M210
- DJI M300, and
- Ascent Aerospace Spirit

The M210 and M300 were found to be the most suitable aircraft for this mission based on considerations that included: (a) flight time between battery changes, (b) availability of applicable sensors, (c) payload capabilities, (d) flight performance characteristics, and (e) landing characteristics (UAV may have to be hand caught).

During these numerous test flights with the BH-12 sensor box – there were no reports of excessive vibration impacting the resulting sensor data.

In port, at anchorage, or transiting within the port

On January 11, 2022, a series of test flights were conducted from a US Coast Guard (USCG) vessel in the Port of Long Beach and outside the breakwater.
Systems tested included – and were not limited to:

(i) Flight Operation Protocol
(ii) Communication Protocol, and
(iii) Aircraft performance.

These flights confirmed the M210 or the M300 to be most suitable for the mission at hand. During these test flights with the BH-12 sensor box – there were no reports of excessive vibration impacting the resulting sensor data.

Tests onboard the Gerner Maersk

On August 10, 2022, numerous flights were conducted on the Gerner Maersk.

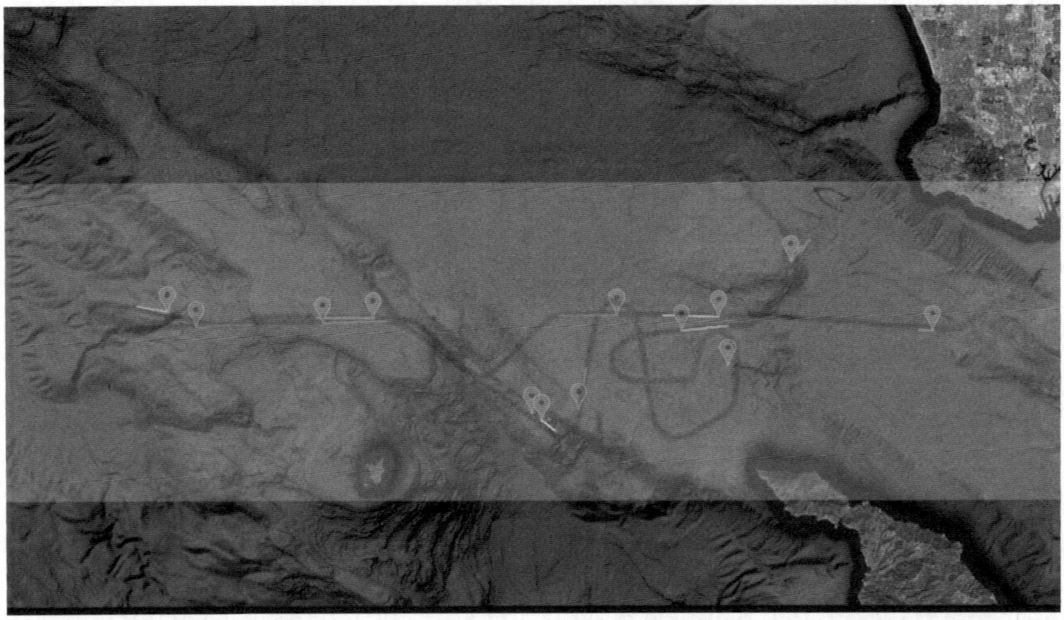

Thankfully - despite, two (2) M210s being determined not to be airworthy due to failure to calibrate the IMU while underway, the M300 could be configured to fly. 13 flights were completed over the course of ~11 hrs. during which 25 samples were captured with the AEROMON BH-12 sensor box. The BH-12 box was observed to vibrate during some of these flights. Attempts to both manually and with the M300s Center of Gravity adjuster were made to minimize such vibrations. Throughout the day there was no mention that any vibrations of the BH-12 box may invalidate the sensor readings. The sole concern was that the BH-12 sensor box may come loose.

Hitachi Portable X-MET 8000 Instrument Methodology

Unlike other compound concentration commercially available techniques (e.g., titration), X-Ray fluorescence (XRF) is rapid, simple and reliable t- and does not require the use of chemicals. These advantages – along with on-going and extensive field use in established industries (e.g., O&G) and in science (e.g., archelogy, geology, etc.) – and California Air Resources Board (CARB) Sulfur Rule calling for the XRF bench method to assess compliance – were deterministic in the inclusion in the research team's test plan for an XRF portable instrument to measure Fuel Sulfur Concentration (FSC) in the fuel samples collected in August 2022 on the Gerner Maersk. Other points favoring this technique were: (i) it is easy to learn and to conduct, and (ii) it is non-destructive, and (iii) the FSC can be read within minutes from when the fuel sample are extracted (vs. the current CARB in-lab bench XRF technique which can take weeks for a sample to be received and analyzed).

Hitachi claimed its field-portable XRF analyzer, the X-MET 8000, combines a high-performance X-ray tube and a large area silicon drift detector (SDD) as to rapidly deliver high accuracy measurements for routine FSC measurements. These claims were verified during this project.

The X-MET is robust (IP54 rated to ensure minimal ingress of dust and splash water and tested to MIL-STD 810G for ruggedness), and is easy to handle (i.e., relatively light weight and compact. The project demonstrated that it can be used on-board a ship, in a laboratory, or wherever the analysis needs to be performed (e.g., in a port terminal parking lot).

The X-MET offers flexible data management. Results are automatically stored in the analyzer and can be transferred to a USB flash drive or a PC, printed on a portable Bluetooth printer, or shared in real-time through the ExTope Connect app and cloud service.

SAMPLE PREPARATION

The below protocol was used to prepare the samples collected on the Gerner Maersk and on the fuel samples collected previously from three MARAD vessels. The fuel is poured into a sample cup fitted on top with thin polyester film (e.g., 3.5μm Mylar®). This cup is placed in the safety window – which is also fitted with thin film. This minimizes the risk for spill, cross-contamination or cup leakage onto the analyzer, and avoids costly repairs. The cup and safety window are placed in the light stand. Pressing the X-MET trigger starts the analysis, and initial results are available in seconds, updating until the end of the analysis.

A typical analysis time to get a marine fuel oil sulfur concentration measurement is 60 seconds.

PERFORMANCE AND RESULTS

The X-MET was calibrated following industry-standard procedures (ASTM D4294): a series of commercially available calibration standards were measured to establish the relationship between sulfur content in the oil and X-ray signal. Each standard was measured for 60 seconds.

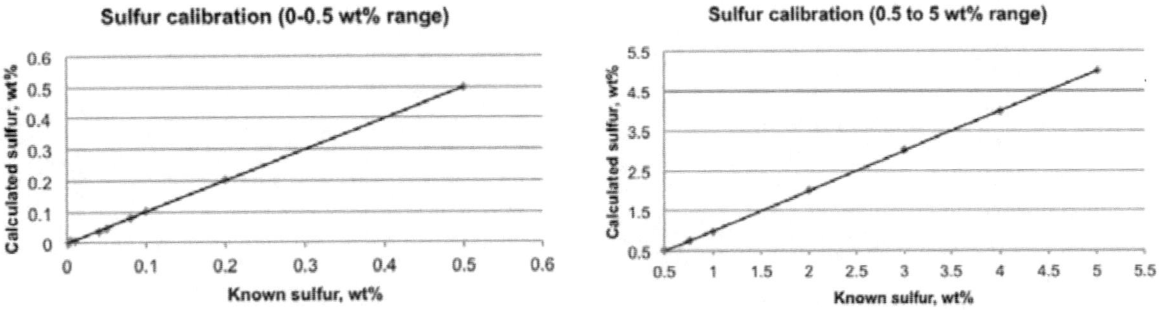

Figure 14. For optimum accuracy across a wide sulfur range (0 to 5 wt%), the calibration is split into 2 ranges. During routine sample analysis, the correct regression is automatically selected, so there is no decision needed from the operator.

Table 1. Certified reference materials (CRMs) were measured to verify the calibration accuracy, and to demonstrate how the X-MET can be used to ensure your fuel meets requirements.

CRMs	Oil type	Given, wt%	X-MET result, wt%
NIST 1623c	Residual fuel oil	0.40	0.38
NIST 1622e	Residual fuel oil	2.12	2.15

SUMMARY

The X-MET's ease of use, ruggedness and portability make it an ideal tool for sulfur analysis on ships, at the ports, or in the lab.

Results are available in minutes - saving time and money.

The X-MET calibration process is simple. Once calibrated - the X-MET provides accurate and repeatable sulfur content analysis in marine fuel oil.

Appendix B: Raw Data

Time (PST)	Hitachi Portable	ISO 8754 (CARB)	ISO 8178 UCR (FS	ISO 8754 (LAB	Aeromon UAV	ISO 10370 (MCRT, wt%)	ISO 10370 (MCRT bottom 10%,	CARB-HITACHI	CARB HITACHI	CARB-LABS D	CARB-LABS D	CARB-AEROMO	CARB-AEROMON DI	CARB-UCR DIF	CARB-UCR DI
6:15	0.0854	0.083	--	0.091	--	--	0.05	0.0024	0.0024	0.008	0.008				
6:30	--	0.0831	--	--	--	7.89	--								
6:45	--	0.0823	0.128439531	--	--	--	--							0.04614	0.04614
7:00	--	0.0801	0.111227999	--	--	--	--							0.031128	0.031128
7:12	--	--	--	--	0.123	--	--								
7:14	--	--	--	--	0.098	--	--								
7:15	--	0.0812	0.102948487	--	0.098	--	--					0.0168	0.0168	0.021748	0.021748
7:17	--	--	--	--	0.122	--	--								
7:20	--	--	--	--		--	--								
7:30	--	0.08	0.096935072	--	0.098	--	--					0.018	0.018	0.016935	0.016935
7:33	--	--	--	--	0.098	--	--								
7:37	--	--	--	--	0.135	--	--								
7:40	--	--	--	--	0.15	--	--								
7:43	--	--	--	--		--	--								
7:45	0.0814	0.0794	0.099642194	0.089	0.15	--	--	0.002	0.002	0.0096	0.0096	0.0706	0.0706	0.020242	0.020242
8:00	--	0.0799	0.086069603	--	--	--	--							0.00617	0.00617
8:15	0.1496	0.1443	0.085571076	--	--	1.24		0.0053	0.0053					-0.058729	0.058729
8:30	--	0.238	0.102254296	--	--	--								-0.135746	0.135746
8:40	--	--	--	--	0.238	--									
8:44	--	--	--	--	0.268	--									
8:45	0.3236	0.3071	0.12442958	0.331	0.268	4.78		0.0165	0.0165	0.0239	0.0239	-0.0391	0.0391	-0.18267	0.18267
8:47	--	--	--	--		--									
9:00	--	0.3429	0.173045355	--	--	--								-0.169855	0.169855
9:15	0.3848	0.3814	0.205151967	0.402	--	6.16		0.0034	0.0034	0.0206	0.0206			-0.176248	0.176248
9:27	--	--	--	--		--									
9:29	--	--	--	--	0.344	--									
9:30	--	0.4009	0.265556567	--	0.344	--						-0.0569	0.0569	-0.135343	0.135343
9:45	0.4307	0.4068	0.350984042	--		7.07		0.0239	0.0239					-0.055816	0.055816
10:00	--	0.4275	0.379038276	0.443	--	--				0.0155	0.0155			-0.048462	0.048462
10:06	--	--	--	--	0.385	--									
10:09	--	--	--	--	0.407	--									
10:12	--	--	--	--	0.389	--									
10:15	0.4384	0.438	0.407607926	0.467	0.358	7.53		0.0004	0.0004	0.029	0.029	-0.08	0.08	-0.030392	0.030392
10:30	--	0.4371	0.423993538	--	--	--								-0.013106	0.013106
10:36	--	--	--	--	0.342	--									
10:39	--	--	--	--		--									
10:42	--	--	--	--		--									
10:45	0.4394	0.4197	0.449375394	--	--	7.17		0.0197	0.0197					0.029675	0.029675
11:00	0.3575	0.3417	0.445803231	0.362	--	--		0.0158	0.0158	0.0203	0.0203			0.104103	0.104103
11:15	0.2826	0.2684	0.426260636	--	--	3.93		0.0142	0.0142					0.157861	0.157861
11:30	--	0.2222	0.400358585	--	--	--								0.178159	0.178159
11:45	0.1932	0.182	0.360183055	--	0.226	2.04		0.0112	0.0112			0.044	0.044	0.178183	0.178183
11:47	--	--	--	--	0.226	--	--								
11:49	--	--	--	--		--	--								
12:00	--	0.1458	0.34928183	--	--	--	--							0.203482	0.203482
12:15	0.1383	0.1299	0.313908575	0.141	--		13.83	0.0084	0.0084	0.0111	0.0111			0.184009	0.184009
12:30	--	0.1157	0.27759636	--	--									0.161896	0.161896
12:45	--	0.1008	0.263010389	--	--		4.48							0.16221	0.16221
13:00	0.0957	0.0982	0.235973835	0.107	--		--	-0.0025	0.0025	0.0088	0.0088			0.137774	0.137774
13:15	--	0.0942	0.208057108	--	--		2.47							0.113857	0.113857
13:30	--	0.0927	0.17539814	--	--		--							0.082698	0.082698
13:45	0.0902	0.0861	0.158141431	0.096	--		1.38	0.0041	0.0041	0.0099	0.0099			0.072041	0.072041
14:00	--	0.0845	0.132108582	--	--		--							0.047609	0.047609
14:15	--	0.0807	0.129045354	--	--		0.66							0.048345	0.048345
14:30	0.0868	0.0815	0.129993318	0.09	--		--	0.0053	0.0053	0.0085	0.0085			0.048493	0.048493
14:45	--	0.0808	0.132446778	--	--		0.52							0.051647	0.051647
15:06	--	0.0809	0.133295051	--	--		0.51							0.052395	0.052395
15:15	0.0838	0.0806	0.158247563	0.09	--		0.34	0.0032	0.0032	0.0094	0.0094			0.077648	0.077648
15:30	--	0.0814	0.150004283	--	--		0.48							0.068604	0.068604
15:45	--	0.0806	0.120064483	--	--		0.27							0.039464	0.039464
16:00	0.083	0.0809	0.1322825	0.09	--		--	0.0021	0.0021	0.0091	0.0091			0.051383	0.051383
16:15	--	0.081	0.168827259		--		0.23			-0.081	0.081			0.087827	0.087827
16:30	--	0.0775	0.161025227		--		--			-0.0775	0.0775			0.083525	0.083525
16:45	--	0.0804	0.142425037		--		--			-0.0804	0.0804			0.062025	0.062025
17:00	0.0828	0.0798	0.141998529		--		0.18	0.003	0.003	-0.0798	0.0798			0.062199	0.062199
								HITACHI		LABS		AEROMON		UCR	
** red entries indicate a copied data point from < 5 min difference for statsitical comparisons to CARB data points								MIN		MIN		MIN		MIN	
								-0.0025	AVG	-0.081	AVG	-0.08	AVG	-0.18267	AVG
								MAX	0.0079667	MAX	0.029553	MAX	0.046485714	MAX	0.087996
								0.0239		0.029		0.0706		0.203482	

READY FOR TAKE-OFF? TOWARDS DETECT AND AVOID FOR COMMERCIAL UAVS

Vidhya Sridhar *

Commercial drones, and the business models they enable, will truly take off when fleets of uncrewed aerial vehicles (UAVs) of many shapes and sizes can operate autonomously in shared airspace. For this to happen, UAVs need to be able to fly beyond the visual line of sight (BVLOS) of remote human pilots, and the technologies that make this possible need to come at an acceptable size, weight, power and cost (SWaP-C).

Here, we focus on the toughest technological challenge on the way to UAV autonomy: last resort tactical airspace deconfliction by means of onboard sensing, also known as Detect and Avoid (DAA). We consider the safety goal posts, what can be achieved with different sensors and data processing techniques, and whether we can meet industry needs for SWaP-C and chart a course towards certification.

INTRODUCTION

From delivering vital medical supplies in regions where other infrastructure is lacking to transporting people in air taxis, UAVs are here to stay. But to move beyond the current waiver-based operations and unlock new business models, UAVs need to be able to fly BVLOS in a safe and scalable manner.

We have previously introduced the different technical puzzle pieces needed to make this happen. Ultimately, the ability to mitigate risk hinges on conflict resolution in shared airspace. However, we, the industry, need to solve this problem in a manner that makes adoption of solutions easy. This implies solutions that not only meet performance requirements but do so within the payload limitations of different UAVs and come at an appropriate cost point.

HOW DO WE AVOID AIRSPACE CONFLICT?

The three layers of airspace conflict resolution are:

- Procedural resolution: plan the operation with sufficient separation from other airspace users;
- Air traffic management: use a third-party service whose subscribers share data to improve mutual awareness and communication with other airspace users; and
- Detect and avoid (DAA): last resort tactical deconfliction.

* Head of Autonomous Technologies, TTP plc, TTP Campus, Cambridge Road, Melbourn, SG8 6HQ, UK

Last resort tactical deconfliction is a crucial but challenging technological hurdle consisting of three separate problems – detection, alerting and avoidance. In this paper, we focus on detection since this remains the bottleneck in terms of meeting SWaP-C and performance.

The market offering for avoidance is relatively mature. Several “avoid” technologies compliant with regulation are available. The alerting function has some interesting challenges for certain use cases such as optionally piloted vehicles (OPVs). What and how should information be displayed to the pilot? There is also interest in learning from heads-up display (HUD) technology in the automotive sector, and how this may transfer to electric vertical take-off and landing (extolls). For the rest of this paper, we focus on detection.

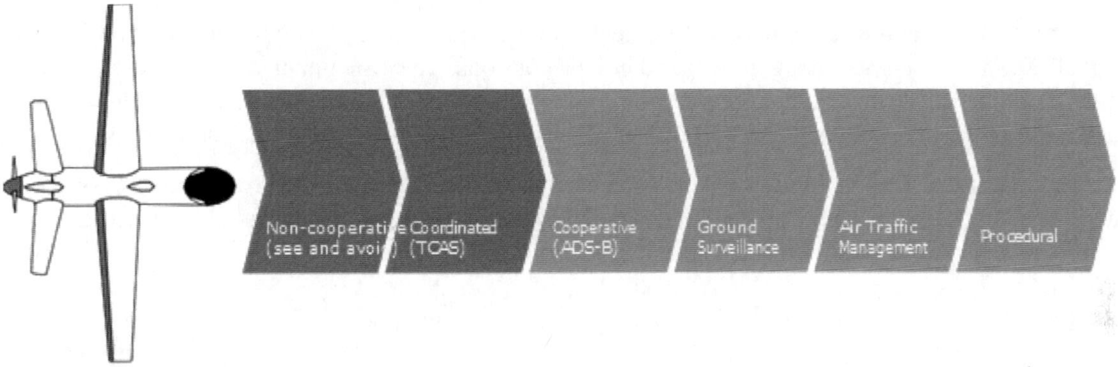

Figure 1. An illustration of the various means to resolve conflict in airspace[1].

First, let’s take a step back and set the safety goal posts.

HOW IS SAFETY DEFINED?

One of the challenges in developing DAA systems is uncertainty about the goal posts. NASA considers statistical metrics (termed as lagging indicators) and performance metrics (termed as leading indicators)[2]. Examples of statistical metrics include safety targets such as number of fatal accidents per flight hour set by the International Civil Aviation Organisation (ICAO), or general aviation mid-air collision statistics. Statistical measures such as these are largely unsuitable as development goal posts because they are not representative of the type of encounters that UAVs will experience and will not scale as the skies fill up with UAVs.

Performance based metrics are more suitable for defining goal posts. They are derived from currently accepted levels of aircraft safety. This is done by (a) defining encounter models of uncrewed-crewed and uncrewed-uncrewed traffic and (b) quantifying the physical and mental limitations of a human pilot. This has now formed the basis of the aircraft collision avoidance standards for small aircraft (ACAS-sXu)[3] published by ASTM International in 2020.

The ACAS sXu specification uses cylindrical boundary boxes around a UAV to define airspace boundaries for 'near mid-air collision' and 'well clear'. The objective of any DAA system is to keep the probability of detecting another airspace user within the cylindrical boundary boxes under a

[1] Adapted from Taxonomy of Conflict Detection and Resolution Approaches for Unmanned Aerial Vehicle in an Integrated Airspace

[2] Sense-and-avoid equivalent level of safety definition for unmanned aircraft systems. revision 9, NASA, 2005

[3] Standard Specification for Detect and Avoid System Performance Requirements, ASTM International, 2020

certain threshold. This, in conjunction with the encounter models, provides a basis for determining the detection range vs. the angular rate accuracy required from a DAA system. Let's look at some examples to get some tangible numbers.

Take a small cargo drone with a maximum take-off weight (MTOW) of about 40 pounds and cruising speed of about 40 knots. Most UAVs in this class are designed to operate autonomously with minimum pilot intervention – i.e., whilst a collision avoidance operation needs supervision from a remote pilot and can be overridden by them, the pilot wouldn't need to execute the manoeuvre. Banking angle and turn rates can depend on whether the aircraft is multi-rotor or fixed-wing. Based on the encounter model statistics and risk ratio requirements in ACAS-sXu, the DAA system would need to achieve a detection range of at least 1.5 km with an angular rate accuracy of 0.7 degrees/ second.

Note that this is the range needed to detect crewed aircraft and does not consider "unconventional" airspace users like hang gliders and hot air balloons. They are unconventional now, but it is not far-fetched that they might become more common and would need to be detectable (albeit a detection range of less than 1.5 km would be acceptable for slower moving airspace users).

The other question is angular coverage. The Federal Aviation Administration insists on 360-degree azimuthal coverage. In other countries, Mexico for example, a front facing coverage, as for a crewed airplane, is considered sufficient.

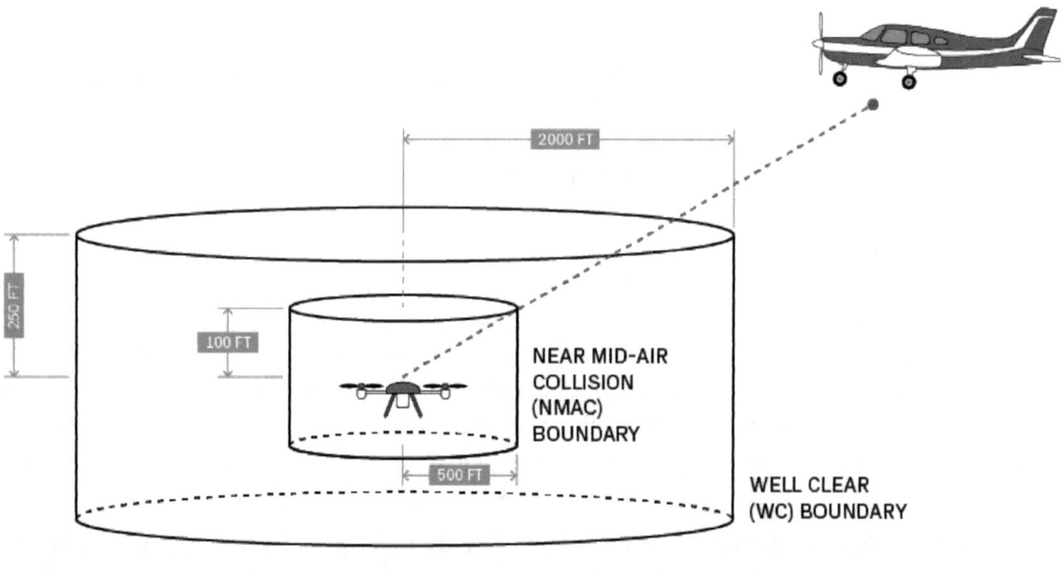

Figure 2. The NMAC and WC boundaries used to benchmark the performance of DAA systems[4].

WHAT DOES THE INDUSTRY NEED?

We talked to stakeholders to find out. Our conversations suggest that the requirements for detection systems vary, depending on use case and MTOW.

---

[4] Adapted from ACAS-sXu specification

At the lower end of MTOW – corresponding to UAVs used for inspection and small cargo delivery (both intra-city and long distance) – companies were very sensitive to SWaP-C. A wish list of weight < 1 kg, power consumption < 10-20 W, cost ~ $1000s, and detection range in the order of 1 to 2 km was typical.

Developers of larger, cargo-carrying UAVs and air taxis, whilst also sensitive to cost and size were less sensitive to weight. However, amongst these, those flying in controlled airspace require airborne solutions with much longer range (2 to 3 km) since the closing speeds would be very high.

WHAT IS POSSIBLE?

The toughest technology challenge is undoubtedly an onboard DAA solution for low MTOW aircraft and any solution would need to push the limits of sensing, deep software, SwaP-C, and performance. Based on our conversations with stakeholders, we set ourselves a target of range of detection of greater than 1 km at an aggressive SWaP-C budget of 500g, 10W and BoM $1000.

We need to resolve:

- How close to that can we get?
- What is possible in terms of performance for different sensing modalities and SwaP-C?
- Are we able to demonstrate a route to certification?

To address these questions, we took a ground-up approach by looking at the minimum sensor specification that met performance needs.

VISUAL TECHNIQUES

Visual approaches are a good candidate for low MTOW UAVs. We have developed and prototyped several architectures and algorithms to estimate what is possible.

One architecture employs a combination of four wide field of view (FOV) cameras to provide 360-degree coverage and a steerable narrow FOV camera. Output from the wide-angle cameras is used to estimate motion and classify pixels as either background or another airspace user. In the video (Video 1), this classification technique is used to detect hang gliders from a single low- resolution GoPro camera. This approach detects and tracks hang gliders with few false positives.

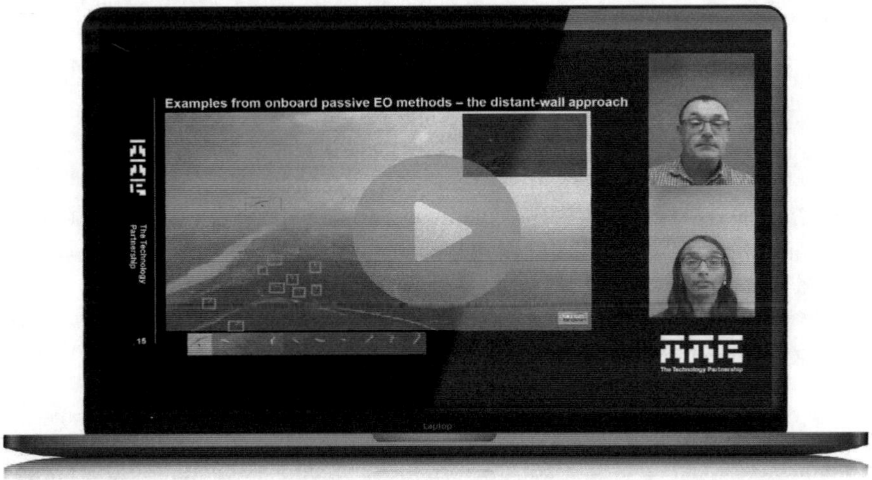

Video 1. Demonstration of approach where estimation of motion is used to classify pixels as background or airspace user.

Other, lower complexity architectures for motion estimation could use the drone's IMU data. The airspace user detection stage is used to steer the narrow FOV camera, the output of which is fed to an object classification algorithm. Video 2 shows an object classification machine learning (ML) model trained on a small dataset of helicopters seen from the ground. Inputs to such a classifier could be low resolution patches from the wide-angle system or high-resolution images from the steerable camera.

Video 2. Demonstration of object classification machine learning trained on a relatively small data set of helicopters seen from the ground.

Another architecture, consisting of multiple cameras covering every possible direction, allows stereoscopic distance measurement in a single image frame and removes the uncertainty introduced by the UAV's own motion estimation.

Our analysis suggests that while vision-based systems may not quite yet meet our SWaP -C targets, they can come close to the required performance at about ~ 1kg and ~15 W, at BoM cost below $1000. This is lower than currently available vision-based approaches.

Vision-based approaches also have a role to play in DAA solutions for higher MTOW UAVs. While they would not be the first line of defence, because they cannot provide the range needed, they would serve as a second line of defence to enhance classification of aircraft.

It is worth mentioning that we had also explored LiDAR as a candidate technology for DAA - however, the complexity and cost for the range needed quickly ruled this out.

WHAT ABOUT ACOUSTIC SYSTEMS?

To work out what a DAA system could achieve with acoustic detection, we went back to first principles and worked out an audio link budget. Based on this, a conservative estimate of the detection range of a light aircraft is ~700 metres. This calculation assumes an acoustic beamformer (32 mics) to reduce uncorrelated environmental and correlated aero acoustic noise. However, this estimate is conservative since it assumes an SNR of 10 dB for detection, which could be further improved by post-processing.

We tested a simplistic principal component analysis (PCA) based post-processing approach on a relatively small, labelled UAV data set comprising drones, helicopters and background noise. This resulted in several false positives, suggesting that more sophisticated data processing approaches may be required. More sophisticated approaches could look at periodic/harmonic structures in the frequency spectra – for example, the figure below shows spectral features that can be obtained for helicopters – in conjunction with machine learning approaches.

Clearly, getting large enough data sets is key to obtaining these processing gains. This also does not include other noise reduction approaches that can be integrated into the airframe. For example, Zipline, who have launched an acoustic DAA solution with their airframe, have explored modifications to the aircraft wing edges to reduce aero-acoustic noise.

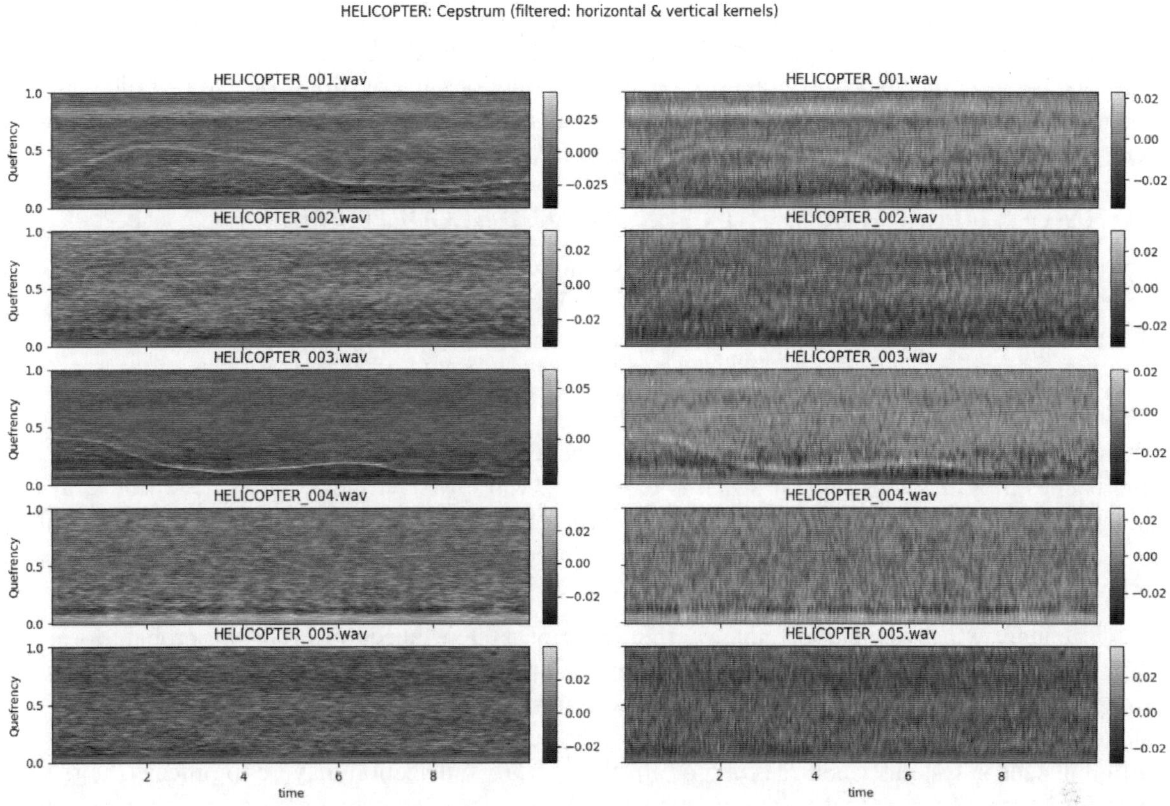

Figure 3. Analysis of audio from helicopters showing specific spectral features in the frequency spectra.

Even assuming that processing gains will result in an improved detection range of 1-2 km, the downside of the acoustic approach is poor angular accuracy. Plus, acoustic approaches simply cannot detect non-conventional users like hang gliders and hot air balloons. The upside is that the SwaP-C cost of acoustic detection is almost free. We could get away with ~ 1W, $150 and size of 600 mm2.

Based on our analysis, we think that passive acoustic methods may serve the market in the short term. In the longer term, acoustic approaches may become a second line of defence in a sensor fusion DAA system for smaller MTOW aircraft.

LAST BUT NOT LEAST: RADAR SYSTEMS

The other approach on the table is radar. K-band radar solutions such as those produced by Echodyne and Honeywell are attempting to capture the DAA radar market. In the K-band, to detect a cross-section equivalent to a Cessna at 1 km would require about 30 dBi gain. To keep the form factor small, this implies Effective Isotropic Radiated Power (EIRP) in the range of 80 to 90 dBm, which would drive up the power requirement to quite a way outside our SWAP-C budget for low MTOW UAVs.

An assessment of mm wave radar – which is not currently a regulated air-to-air radar band – suggests we could obtain a 2 km range with a ~30 cm antenna aperture, albeit with a high-power requirement. Overall, for this end of the market, a pure radar-based solution seems unlikely whilst a tip and cue-based sensor fusion approach might be where radar could have a place.

However, radar is certainly the way to go for large UAVs. K-band and X-band radar solutions can provide the range needed, given that the power requirements can be easily met by larger UAVs. The challenge for the market would be to obtain these solutions at a sensible price point which may require pairing up off-the-shelf radar core modules and suitable off-the-shelf (OTS) or custom antennas that can meet size, drag and coverage requirements.

COPING WITH IMPAIRMENTS IN SENSING AND ALGORITHMS

The discussion so far highlights the strengths and weaknesses of different sensing technologies and their role in DAA for low and high MTOW UAVs. The table below is a summary and flags up other environmental disturbances that can impair the sensing system.

This shows that no single sensing technology will meet the performance needs in all environments, which suggests that sensor fusion may become more important in future. Furthermore, the choice of primary sensing modality may differ depending on MTOW. For example, vision-based approaches may be more suitable for smaller UAVs, while RADAR may be more so for larger UAVs. Secondary sensing systems, beyond improving performance of the primary system, will more importantly provide minimum fallback performance in environmental conditions where the primary sensor fails to perform.

Ultimately, whatever the technology, the detection system needs to drive the probability of "false negatives", the probability of missing a potential airspace threat, as low as possible. Whilst we would also like to keep the probability of "false positives" low, i.e., misclassifying a non-threat as a threat, this probability can be traded off somewhat against false negatives. Apart from using different sensor systems, another prong of attack to improve detection may be to process the raw data output from any one sensing system with two or more complementary algorithmic approaches.

Table 1. Summary of pros, cons and suitability of various sensor systems for different classes of UAVs.

Sensor technology	Low MTOW		High MTOW	
	HIGHS	LOWS	HIGHS	LOWS
Vision	▪ Good range (at low closing speeds) ▪ Good accuracy ▪ Reasonable SWaP-C	▪ No night and poor weather operation ▪ Will need to address specular reflections	▪ Can aid airspace user classification	▪ Poor range (at high closing speeds) ▪ No night and poor weather operation
Acoustic	▪ Extremely low SWaP-C ▪ Long range possible (with several caveats around ability to cope with noise)	▪ Poor angular accuracy ▪ Cannot detect "quiet" airspace users (like hang gliders)	NOT SUITABLE	
Radar	▪ Good detection range ▪ Good angular accuracy (traded off against range at this MTOW)	▪ High SWaP-C ▪ Will need to address ground clutter ▪ Will need to address specular reflections	▪ Good range ▪ Good angular accuracy ▪ Suitable SWaP-C	▪ Performance impacted by clutter ▪ Will need to address specular reflections

HOW WILL WE CERTIFY THESE SYSTEMS?

For any DAA system, especially one using passive methods and artificial intelligence techniques, how will we demonstrate safety? Some existing solutions rely on the guarantee of many flight hours without incident. However, this will never cover all possible corner cases. Aviation regulators are starting to recognise that there is an important place for synthetic data and simulation tools that can extend data coverage and reduce costs incurred in collecting flight data.

With synthetic data come the added burdens of ensuring that simulation adequately matches reality and of developing means to verify the integrity of the data. This is certainly achievable – making it tractable requires an intelligent assessment of which variables are correlated and uncorrelated. Nevertheless, it will never be possible to test a DAA system against all possible scenarios that will be encountered in the air. Thus, it is also important to understand the limitations of the system.

Traditional development assurance frameworks in aviation do not cover machine learning approaches well, i.e., they are designed for situations where "each line of code can be verified". Now the industry is considering ways to perform rigorous data and inference verification and validation as part of the software certification process. Beyond that, we could ensure our software architecture lends itself more easily to verification and certification. Clearly, an architecture that modifies the neural network "on the fly" based on data observed in flight will face an uphill battle for certification. An architecture that relies on classical methods where possible and neural networks with well-bounded behaviour and training data would be a more balanced approach that could be acceptable by the regulator.

COME AND TALK TO US!

With our first level analysis out of the way, we now have our arms well around the DAA challenge. We are now exploring partnerships to develop the next generation of DAA systems. Depending on your market and what you are looking to achieve, be it lower SWAP, better performance and/or a clearer path to certification, there are various ways TTP can work with you:

- If you are a DAA solutions provider, TTP can assist in the development of your next-generation system. Have you considered sensor fusion, or another ground-up approach using commercial high-volume sensors to minimize SWAP?
- If you are a regulator, TTP – as an independent technology company – can work with you to help develop a staggered technology approach towards certification.
- If you are a UAV operator looking to integrate, customize or develop a suitable DAA solution, have a chat with TTP's autonomous technology experts about your airframe and operational constraints to identify next steps on your DAA journey.

Or is there perhaps a different technical challenge that could benefit from TTP's expertise, scale and agility? Want to know more about our journey towards Detect and Avoid for Commercial UAVs?

enquiries@ttp.com

CHALLENGES AND PROGRESS IN BEHAVIOR-BASED ADAPTIVE AUTONOMY

Arjuna Balasuriya,[*] Tyler Mayer,[†] and Gerald Fry[‡]

Today's uncrewed platforms are typically operated by humans using remote control to guide every detailed aspect of a mission. However, as missions become more complex, there are many scenarios (particularly in the marine, ground, and space domains) in which operators are unable to communicate with these uncrewed platforms in real time (due to adverse environmental conditions, regulatory restrictions on communications in ecologically sensitive areas, active interference by adversaries, or the desire to remain covert), making it challenging to recalibrate and update mission and control parameters on the fly. Fortunately, significant technical advances in onboard computing power and enhanced sensors offer a pathway to a level of autonomy that can overcome such communications limitations. Behavior-based autonomy architectures enable onboard mission autonomy software to select optimal behaviors for the mission at hand, assess the evolving state of the mission, and adapt to changing environmental conditions. Developers and operators of autonomy systems for challenging environments will gain insights into the importance of onboard behavior and parameter adaptation to the success of long-duration uncrewed platform missions in unpredictable situations and unknown environments. Designers will benefit from a discussion of how multi-objective optimization and reinforcement learning techniques are currently being used to enable assured autonomy onboard these uncrewed platforms.

INTRODUCTION

Uncrewed platforms have become increasingly popular in recent years and have been employed for various commercial (e.g. household appliances, self-driving cars, search and rescue operations, manufacturing, agriculture, and uncrewed space travel) and defense applications (e.g., uncrewed platoons, intelligence, surveillance and reconnaissance (ISR), and missile defense) across space, air, land, underground, sea surface, and underwater domains.[1] Currently, most of the uncrewed platform operations are carried out in structured or known environments with the human in the loop. For example, uncrewed underwater vehicles (UUVs) are widely used for surveillance and mapping applications. UUV operators plan missions and calibrate the sensors, by studying the bathymetry and environmental properties, such as conductivity, temperature, and depth in the operating environment. In most of the current operations, UUV operators track the UUV's position

[*] Senior Scientist, Sensing Perception and Applied Robotics Division, Charles River Analytics, 625 Mt. Auburn St. Cambridge, MA 02138.

[†] Scientist, Human-Centered AI Division, Charles River Analytics, 625 Mt. Auburn St. Cambridge, MA 02138.

[‡] Scientist, Human-Centered AI Division, Charles River Analytics, 625 Mt. Auburn St. Cambridge, MA 02138.

while it is executing its mission underwater and have emergency abort devices to recover the vehicle in case of any unexpected changes to the environment or mission objectives. The real benefits of uncrewed platforms can be realized only if these platforms are able to adapt to changes in the environment and mission objectives with minimum operator intervention. These platforms can operate in unknown, highly complex, and dynamic environments only by improving their onboard autonomy algorithms. Operators must be able to trust these autonomy algorithms.[2] This becomes a challenge due to unreliable sensors and harsh environmental conditions. For example, acoustics sensors are the most effective option for underwater use; however, their performance is degraded due to continuous changes in acoustic properties in the operating environment. Similarly, continuous changes in water density due to mixing and underwater currents make it hard for these platforms to operate effectively in these highly dynamic environments. As a result, there is a high demand for autonomy algorithms to include resource management, maintenance, and forecasting algorithms for these platforms to operate reliably and interact with human operators only when they need assistance. This paper looks at some of the autonomy frameworks commonly used for uncrewed platforms in highly dynamic environments and discusses how onboard autonomy can be made adaptable/trustable to uncertain sensors, unexpected faults to the platform (e.g., hardware failure or software resource limitations such as battery, memory, or processing power), environmental changes, and changes to mission objectives.

We will present some of the commonly used behavior-based autonomy frameworks and then discuss in the following sections how these frameworks are been implemented in uncrewed platforms to make them adaptive to any unexpected internal and external changes.

REVIEW OF BEHAVIOR-BASED AUTONOMY FRAMEWORKS

An autonomous system should be able to select among multiple possible action sequences to successfully achieve its goals. Behavior-based autonomous system research has been inspired by biological organisms, with the aim of understanding and implementing basic, survival-related behaviors in uncrewed platforms, and advanced behaviors involving, for example, high-level reasoning. Most of the behavior-based autonomy architectures are layered in such a way that the capabilities of the uncrewed platform are broken down into individual behaviors, as illustrated in Figure 1.

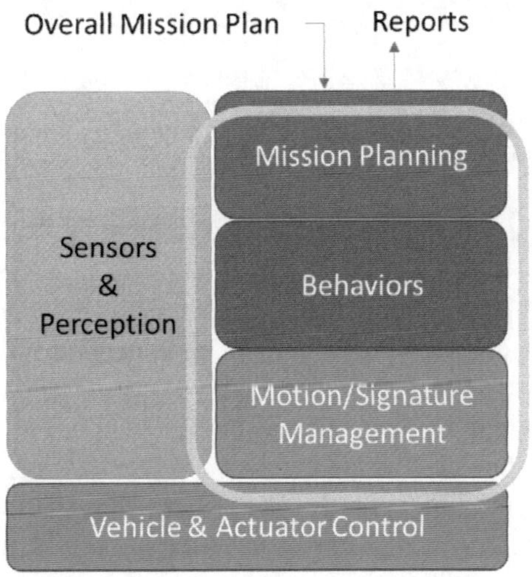

Figure 1: Behavior-Based Layered Architecture Onboard an Uncrewed Platform.

As shown in Figure 1, sensors mounted on an uncrewed platform and its perception subsystem will create situational awareness (i.e., internal (self-awareness) and external (state of the environment) states). The mission planning subsystem will then analyze the situation with respect to the overall mission at hand to determine the current mission objectives. The mission planning subsystem will activate behaviors to achieve the current mission objectives. Each active behavior will determine the best course of action for the given situation and mission objective. However, all the behaviors will be competing for resources onboard the uncrewed platform and the motion/signature management system needs to identify the best action for each moment and command the vehicle and actuator control systems. An example of a layered autonomy framework used onboard a UUV is shown in Figure 2.

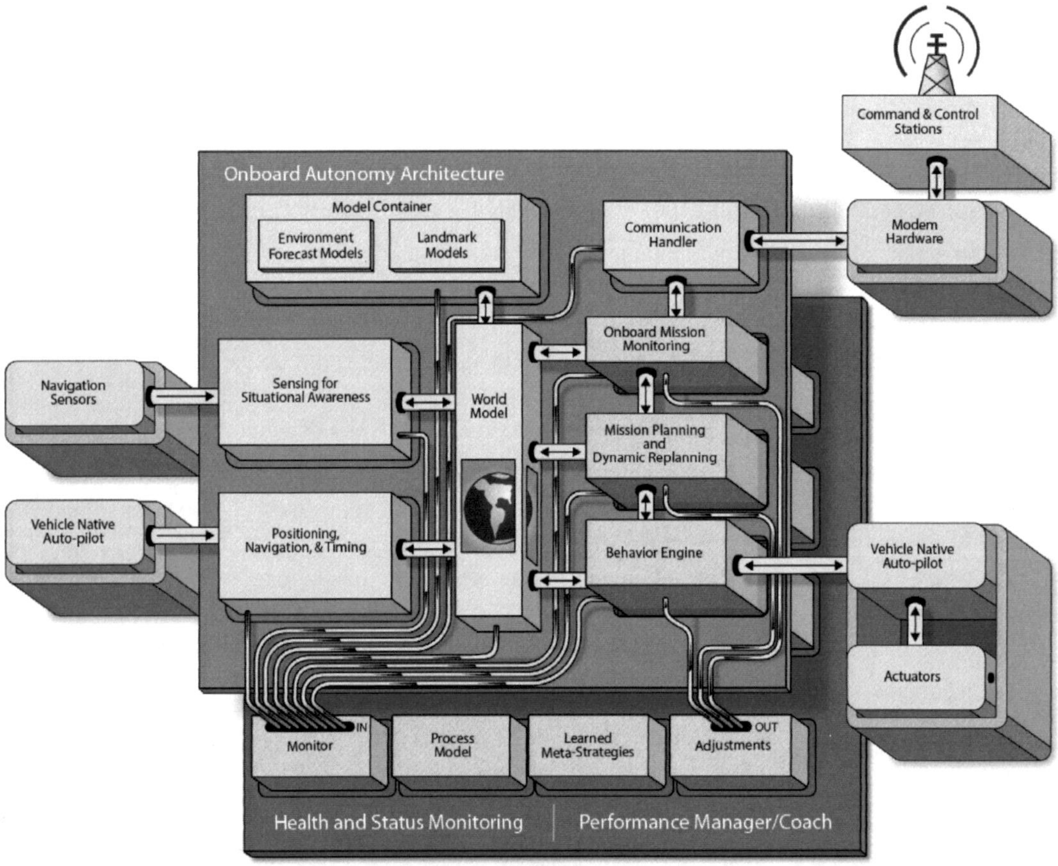

Figure 2: An Example of a Behavior-Based Autonomy Architecture.

A typical UUV is fitted with internal–external situational awareness sensors and position, navigation, and timing (PNT) sensors such as inertial sensors, GPS, and compass. Commonly used sensors include internal environmental sensors for health and internal status monitoring of the following: pressure, leak, humidity, temperature, pressure, and current/voltage sensors. External sensors monitor a range of factors, conditions, and equipment, such as conductivity, temperature, depth (CTD); Doppler velocity log (DVL); sonar (active–passive); camera; pressure; and water current. As shown in Figure 2, UUVs run numerical environmental forecasting models to keep track of the environmental changes. For example, *reachability front behavior* will keep track of the underwater currents to optimize the path to its destination, as shown in Figure 3.

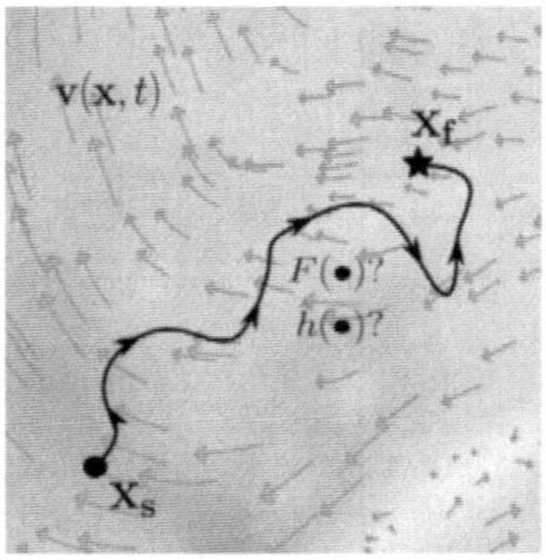

Figure 3: Time-Optimum Path Calculation to Travel from X_s to X_f under Ocean Currents v(x,t).[3]

UUVs need to be conscious about the changes in the environment. For example, due to the range and attenuations, sound waves are the most successful medium used underwater for communication and sensing (lessons learned from the biological systems). However, underwater acoustic properties are constantly changing, which impacts the waveguide properties, and it is important that UUVs are aware of these changes to sense the environment and communicate with the surface or other UUVs, as illustrated in Figure 4.

(a) *ThermoclineDetection* behavior (b) *SurfaceCommunication* behavior

Figure 4: Example of an Adaptive Behavior Used by a UUV to Establish Successful Acoustic Communications with Surface Platforms.[4]

Figure 4 illustrate how a UUV will move in the water column ("yo-yo" motion) to identify the temperature gradient; the region with rapid decrease in temperature over depth is called the thermocline. The speed of sound changes with the temperature and a thermocline may act as a wall blocking any acoustic messages; thus, it is important that the UUV be positioned above the thermocline if it wants to communicate with a surface platform (e.g., ship). Other commonly used UUV behaviors are listed in Table 1.

Table 1: Examples of Common UUV behaviors

Native Behaviors	Description
ConstantDepth	UUV maintains a constant depth
ConstantAltitude	UUV maintains a constant height above the seafloor
WayPoint	UUV travels to its next waypoint
OpRegionBounce	UUV stays inside its operation region
Loiter	UUV loiters around a waypoint
ObstacleAvoidance	UUV avoids obstacles in front of the vehicle
YoYo	UUV changes depth between two levels
DetectThremocline	UUV detects the thermal gradient of the water column
TimeOptimum	UUV follows the optimum path generated by the reachability front

In this paper, we discuss two challenges faced by behavior-based adaptive autonomy systems:

1. Multi-objective Optimization (MOO): As explained above, multiple behaviors active onboard uncrewed platforms will be competing for onboard resources (actuators, processing power, battery power, etc.), and it is important to select the best action (optimum) for a given situation to successfully achieve mission objectives.

2. Performance management: Due to unexpected environmental changes or faulty hardware, the performance of the uncrewed platform can change during mission execution. For example, the power budget of an uncrewed aerial platform can change due to heavy wind conditions.

In the next section, we will present some of the popular MOO and performance algorithms used by uncrewed platforms to identify the best action for unexpected situational changes.

CHALLENGES IN BEHAVIOR-BASED ADAPTIVE AUTONOMY

There are many challenges for uncrewed platforms operating in unknown environments for a long period of time without human intervention. As explained earlier, uncrewed platforms can navigate highly complicated environments for exploration or for executing various tasks, keeping humans away from these unfriendly environments (e.g., deep-ocean environments, space, disaster areas, contested environments, etc.).

Challenge 1: Communication. In many applications, there is no or limited connectivity to the uncrewed platform from the command and control (C2) stations. Therefore, it is necessary that the onboard autonomy can adapt to the unexpected changes in the environment as well as for any changes within the uncrewed platform.

Challenge 2: Sensor uncertainties. The situational awareness (operating environment and the internal state of the platform) of an uncrewed platform is determined by the perception subsystem using various sensors. However, due to complex environmental conditions, sensors can be noisy. For example, sonar sensors used underwater will have clutter due to the changes in the acoustic properties of the environment as well as other factors such as multipath, and signal attenuations, etc. These uncertainties need to be considered when reacting to the determined situation.

Challenge 3: Hardware constraints. Due to harsh environmental conditions, there is a higher likelihood of the hardware's performance deteriorating, and the autonomy system need to constantly monitor the health of the hardware systems.

Challenge 4: Highly dynamic environment. As discuss earlier, uncrewed platforms are often deployed into highly complex environments where severe weather and environmental conditions can create unexpected changes to the operating conditions. The autonomy system should be aware of its own capabilities and be able to adapt its mission plans to unexpected environmental changes. For example, a mission plan might direct the UUV in a certain direction; however, due to heavy underwater currents, even with full forward thrust, the UUV might find itself moving backward. The autonomy system should understand the situation and adapt the plan to change direction and follow a reachability front.

Challenge 5: Limited power budget. Uncrewed platforms are battery powered and missions are planned for a known power budget. However, due to the complex nature of the environment, power consumption may be higher than expected, making it challenging for the platform to complete its tasks. Autonomy systems should keep track of their power availability and adapt their mission plans accordingly.

Challenge 6: GPS denied or unreliable PNT. Knowing the absolute position of the platform is important to navigating in the operating envelope. However, there are many operating domains where GPS is not usable or not available (e.g., underwater). Most of these mobile uncrewed platform navigation decisions will depend on the accuracy of the vehicle's navigation system. The uncertainties associated with the PNT system need to be considered when making mission autonomy decisions.

There are many groups who have worked on developing autonomy frameworks and have implemented them in various types of the uncrewed platforms for various applications. Some of the popular layered autonomy architectures are MIT Mission Oriented Operating Suite – Interval programming (moos-ivp), JPL Control Architecture for Robotic Agent Command and Sensing (CaRACAS), Draper Maritime Open Architecture Autonomy (MOAA), JPL CASPER/ASPEN, CMU UC, etc. These frameworks are effective for static, known environments and for short (on the order of a few days) mission periods. The following sections discuss how MOO and learned meta-strategies can be used to help uncrewed platforms adapt to and handle unexpected situations during mission execution with little or no human interference.

MULTI-OBJECTIVE OPTIMIZATION

In this section we will provide an introduction and motivation for MOO in mission planning, and we will give a brief survey of the state-of-the-art approaches for multi-objective mission planning,

Introduction to MO Mission Planning

In practice there are many competing objectives that describe "success" in a mission. For example, the mission should ideally be completed in a short amount of time, minimize the amount of battery consumption, minimize risk exposure over time, maximize the amount of useful information gained or the value of the kinetic effect deployed, etc. Even in relatively simplistic environments these objectives are often at odds with one another. Consider for example a mission where the goal is to get from point A to B quickly *and* avoid detection by an adversarial vessel. Minimizing the amount of time required to get from A to B may incur great risk (e.g., if the adversary were situated directly between A and B), while minimizing or eliminating risk may require the vessel to get from A to B using a very indirect and therefore suboptimal (with respect to time) path.

In general, there is no single solution (mission plan) that simultaneously optimizes all objectives. A solution is called Pareto efficient if value of any objective cannot be improved without decreasing the value on some other objective (put another way, a solution is Pareto efficient if no other solution achieves a better value on every objective).

The goal of multi-objective optimization is to compute (all or some of) the Pareto-efficient solutions for a given multi-objective optimization problem, and/or the trade-off curves between arbitrary sets of objectives. This requires formalizing the problem as an optimization model and defining a solution space, data (for training, if applicable,) metrics, objectives, constraints, and an algorithm (or policy) for solving that model.

With no other context or subjective preferences across the objectives (e.g., risk tolerance of the mission) all Pareto-efficient solutions are considered equally good. Therefore Pareto-efficient solutions, or trade-off curves, are then fed to the upstream decision-maker, who applies their own instance-specific preferences in determining which solution to pursue. This last step is supported by the broader field of multicriteria decision analytics, which is concerned with studying the positive and negative consequences, risks, and opportunity costs of decision alternatives and helping a decision-maker quantify their subjective preferences when making analytically backed and justified decisions. In our work, and in most of the works surveyed in this section, we assume that we are either providing the Pareto-efficient set to a human operator to make the ultimate decision, or that the human operator has defined some mission-specific preferences over the objectives to facilitate autonomous decision-making.

Survey of Approaches to MO Mission Planning

There is a vast amount of literature in the space of multi-objective mission planning in autonomous systems. We will briefly summarize several popular types of approaches in each of the model-driven and data-driven artificial intelligence (AI) categories.

First, in classic model-based approaches, the goal is to encode relevant aspects of the host platform and the environment, along with their interactions, in a single mathematical model (a mathematical optimization problem) and develop algorithms to cleverly search over the space of feasible solutions to quickly find high-quality solutions. Many works develop novel evolutionary or genetic algorithms[5–7] for single- and multi-UXV mission planning problems. These consider objectives like task allocation, timeliness of mission completion, collision avoidance, etc. Evolutionary and genetic algorithms are a form of search that propose novel solutions by performing some novel kind of mutation to older seed solutions and mimic the process of natural selection.

Wu et al.[8] model the UXV mission planning problem as a path selection problem on a weighted graph. The weights on the graph are objective scores associated with including a vertex or edge in the final solution and consider collision avoidance, corridors, rules of visual navigation, timeliness, and fuel constraints. They apply scalarization techniques to turn the multi-objective problem into a single-objective problem and use single-objective A* algorithm variants for efficiency. This approach is particularly robust as graphs are a fully expressive encoding scheme (i.e., one can encode a wide range of mission planning problems), and single-objective path planning algorithms are extremely efficient. Our approach to Management of Intelligent Navigation for Condition-based Ocean Safe Transit (MIN-COST) was directly inspired by this kind of approach. We take advantage of the efficiency of single-objective path planning problems to build an architecture that automatically searches objective weighting schemes to identify (in the single-objective scalarized space) high-quality solutions to the multi-objective problem. We postprocess the result after the search is completed.

Other approaches consider multitiered architectures that break the overall mission planning problem down into multiple optimal estimation and planning problems to come up with robust solutions in dynamic environments. These include modeling power systems and predicting energy consumption over time to feed in more aureate estimates to the energy optimization loop handled by evolutionary algorithms (among others).[9,10] Ding et al.[11] consider a stochastic hierarchical planning representation of mission planning to forecast motion while trading off mission objectives in cluttered environments with dynamic events. They show that stochastic hierarchical planning operates robustly in uncertain environments and supports contingency management in real time.

In data-driven AI the goal is to train a reinforcement learning (RL) agent to learn a continuous action policy that produces good behavior across a variety of situations. This requires building a reward function (usually within a simulation or labeled data set) that provides a multi-objective reward signal to an agent given an arbitrary vehicle and environmental state. Using reinforcement learning techniques, an agent is programmed to interact with the environment to try and accomplish its task and achieve an appropriately balanced score across objectives. After sufficient training, an agent learns successful sequences of plan fragments as a function of the state. See Tang et al.[12] for a survey of RL approaches and state-of-the-art models.

LEARNED META-STRATEGIES

Various meta-strategies based on machine learning can be effective for enabling autonomous systems to quickly adapt to changes in the environment. In this section, we discuss performance management in general and within the context of an approach to adapting the mission plan of an autonomous UUV to conserve power after an unplanned energy drain (e.g., due to battery failure).

Performance Management for Autonomous Systems

Autonomous systems can be complex with many different components (e.g., path planner, localization, and communication components in a UUV) acting together to achieve mission goals. An approach to performance optimization of the overall system depends upon optimizing the performance of each individual component and their interactions against the mission objectives. To do so, it is helpful to define the *intent* of each system component within the mission context. There are various ways to define intent, as investigated under DARPA.[13] Here, we focus on an approach that defines a set of *intent metrics* and *thresholds* for each system component.

Intent, for a particular component implementation, can be defined either during the development process or after the fact, but requires knowledge about the semantics of the component within a particular mission context. An intent metric for a component is a quantity that can be measured by the monitoring infrastructure of the autonomous platform. These can be continuous measurements or measurements taken after a certain task is completed. For example, an intent metric for an inertial localization component on a UUV platform would be the localization error measured when this information is available (e.g., when the vehicle moves closer to the surface and can access GPS). Another intent metric for the localization component could be an estimated error value (e.g., computed via a Kalman filter) that is computed periodically. An additional example is a waypoint planning component whose intent metric is proportional to the energy consumption resulting from traversing each waypoint. The power usage could be measured periodically during traversal, once reaching each waypoint, or after the vehicle has traversed all waypoints and returned home.

Once intent metrics are defined, intent thresholds can then be applied to them based on mission parameters and available resources. In the localization component example, the mission may require that the estimated and/or the ground truth localization error be maintained under a specified value; otherwise, an *intent violation* occurs. Similarly, an intent violation would be signaled if the sequence of paths specified by the waypoint planner consumes more than a specified amount of energy per unit time. An intent violation signals a system or component-level optimizer that will then take appropriate actions (e.g., adjust navigational/localization parameters, replan waypoints based on different constraints) with the goal of restoring the intent metric values to mission-acceptable levels.

Given intent metrics and their associated thresholds for a collection of system components, an *intent optimizer*, acting autonomously, adjusts the parameters of each component in the system upon each intent violation to bring metric values back within mission requirements. The decision of exactly how to modify component-local and system-wide parameters that modify and control the behaviors of the autonomous system is a complex one, since there could be many interdependencies on the operations and performance of different components within the system. For example, increasing the sampling rate of a sensor filter could increase the power draw due to increased computational load, causing an intent violation for a waypoint planner that attempts to minimize energy use during a mission. These cascading intent violations could be resolved through organizing the system into a component hierarchy defining the *performance* dependencies between settings for component parameters and imposing an ordering on parameter adjustments that will ensure that all intent violations can be resolved without entering an unstable state (e.g., forever "ping-ponging" between intent violations for different components). Therefore, in complex systems with potentially circular performance (i.e., intent) dependencies between component configurations, another approach is needed that can learn these dependencies and enforce higher-level policies on combinations of parameters across all system components.

Machine Learning Approaches to Performance Management

To autonomously and dynamically coordinate adaptation of system component parameters to satisfy intent as described above, we use machine learning techniques based on deep neural networks and reinforcement learning. These techniques can be applied either on a per-component basis (i.e., for adjusting the parameters of a waypoint planner to ensure that the vehicle can return home safely without running out of energy, given the current power profile), or on a system-wide basis if there are nontrivial performance dependencies between component configurations.

Under the Building Resource Adaptive Software Systems (BRASS) program, sponsored by DARPA, we investigated optimizing power usage against objectives for a mission in which an autonomous UUV must search a region of the seafloor for an object. In our experiments, we managed the performance of two different components in a simulation of a UUV system and its environment: (1) a waypoint planner (referred to as a path planner in the reference) that constructs waypoints while executing a "lawnmower" pattern that provides a level of coverage over the search area and (2) a Kalman filter implementation that filters localization sensor readings. These components work together to navigate the vehicle within the search area. We assume the vehicle is equipped with a vision sensor (e.g., camera) that can detect the object if it is within the sensor's limited field of view. The system also includes a capability to monitor the available energy. The primary mission goal is to find the object and return home safely without running out of energy. A secondary goal, if the primary goal cannot be met, is to return home safely despite not finding the object. If the vehicle runs out of energy and cannot return, then the mission fails. To perturb the state of the system, we simulate a battery failure rate that reduces the remaining energy by a random proportion after a random amount of time has passed form the start of the mission. In practice, this immediate energy drain could be a result of environmental factors, such as heat/pressure changes, damage caused by collisions with obstacles/debris, or unexpected power draw from faulty hardware.

To optimize performance to meet the mission objectives, we monitor an intent metric of the waypoint planner that measures the current energy available and compares the results to the expected energy needed to traverse the current waypoint plan. When an intent violation occurs (i.e., the energy needed is greater than the energy available), we adjust the inputs to the waypoint planner that control the shape of the lawnmower pattern that the vehicle will navigate. Note that, in our experiments, we used an automated approach to transform the original code of the waypoint planner to expose new parameters that enabled us to further configure the component, but without semantic knowledge of the direct effects of these new parameters. In many cases, however, components are designed to be highly configurable, and this step is not needed. Similarly, we monitor intent for the Kalman filter component, and perform the same adjustments of parameters to ensure that localization information remains accurate to a given degree.

Upon discovering the intent violation of the waypoint planner, we use a multilayer feed-forward neural network, which we trained on a variety of scenarios within the mission context (e.g., different search regions, search object locations, power perturbations, water currents), to determine the appropriate parameters to the waypoint and Kalman filter components, based on the observed state of the system and the environment (e.g., power available/required, current estimated position error, ground truth position if available at periodic intervals). We then pass the parameters output by the neural network into the waypoint planner, which computes a new route. The new parameters, in our experiments, increased the distance between legs of the lawnmower pattern to reduce the total path length and thus the power necessary to traverse the search area and return home, at the expense of coverage of the vision sensor over the search area. This reduced the probability that the object would be found but would attempt to ensure that the vehicle could return home and not be lost (i.e., mission failure). In our analysis we differentiate between a mission "pass" (finding the object and returning home) and a "degraded" (returning without finding the object) outcome, but we do not distinguish between failure ("failed") cases of finding the object and running out of energy, and not finding the object and running out of energy. Our results show that using our intent-based approach to performance management resulted in about the same number of pass cases (over a variety of mission instances) compared to no adaptation. However, with intent-based performance management enabled, the vehicle was able to return (pass or degraded) in almost 1.5 times more scenario instances. For additional details on our approach and analysis, including results of adapting the Kalman filter component, see Fry et al.[14] and Pfeffer et al.[15]

CONCLUSION

In this paper, we discussed challenges faced by uncrewed platforms for operations in unknown environments with unexpected situations with little or no human operator interference. We reviewed some of the solutions used to make behavior-based autonomy architectures adaptive to overcome these challenges. We discussed how multi-objective optimization algorithms and learned meta-strategies can be used to successfully complete missions by modifying mission and behavior parameters according to the unexpected changes (internal or external to the platform).

REFERENCES

1 AUVSI. (2023, March). *Unmanned Systems & Robotics Database.* http://robotdirectory.auvsi.org/home

2 Edmonds, E. (2019, March). Three in four Americans remain afraid of fully self-driving vehicles. *AAA Newsroom.* https://newsroom.aaa.com/2019/03/americans-fear-self-driving-cars-survey/

3 Subramani, D. N., & Lermusiaux, P. F. (2016). Energy-optimal path planning by stochastic dynamically orthogonal level-set optimization. *Ocean Modelling, 100*, 57–77.

[4] Petillo, S., Balasuriya, A., & Schmidt, H. (2010). *Autonomous adaptive environmental assessment and feature tracking via autonomous underwater vehicles. OCEANS'10 IEEE Sydney*, IEEE, 1–9.

[5] Lamont, G. B., Slear, J. N., & Melendez, K. (2007). UAV swarm mission planning and routing using multi-objective evolutionary algorithms. *2007 IEEE Symposium on Computational Intelligence in Multi-Criteria Decision-Making*, IEEE, 10–20.

[6] Peng, C., & Qiu, S. (2022). A decomposition-based constrained multi-objective evolutionary algorithm with a local infeasibility utilization mechanism for UAV path planning. *Applied Soft Computing*, *118*, 108495. https://doi.org/10.1016/j.asoc.2022.108495

[7] Ramirez-Atencia, C., Bello-Orgaz, G., R-Moreno, M. D., & Camacho, D. (2017). Solving complex multi-UAV mission planning problems using multi-objective genetic algorithms. *Soft Computing*, *21*(17), 4883–4900. https://doi.org/10.1007/s00500-016-2376-7

[8] Wu, P. P.-Y., Campbell, D., & Merz, T. (2009). On-board multi-objective mission planning for unmanned aerial vehicles. *2009 IEEE Aerospace Conference*, IEEE, 1–10.

[9] Liu, J., Wang, W., Li, X., Wang, T., Bai, S., & Yanfeng, W. (2018). Solving a multi-objective mission planning problem for UAV swarms with an improved NSGA-III algorithm. *International Journal of Computational Intelligence Systems*, *11*(1), 1067–1081.

[10] Thompson, F., & Galeazzi, R. (2020). Robust mission planning for autonomous marine vehicle fleets. *Robotics and Autonomous Systems*, *124*, 103404.

[11] Ding, X. D., Englot, B., Pinto, A., Speranzon, A., & Surana, A. (2014). Hierarchical multi-objective planning: From mission specifications to contingency management. *2014 IEEE International Conference on Robotics and Automation (ICRA)*, IEEE, 3735–3742.

[12] Tang, Y., Zhao, C., Wang, J., Zhang, C., Sun, Q., Zheng, W. X., Du, W., Qian, F., & Kurths, J. (2022). Perception and navigation in autonomous systems in the era of learning: A survey. *IEEE Transactions on Neural Networks and Learning Systems*, https://doi.org/10.1109/TNNLS.2022.3167688

[13] DARPA. (2015). *DARPA BRASS Program*. https://www.darpa.mil/program/building-resource-adaptive-software-systems

[14] Fry, G., Samawi, T., Lu, K., Pfeffer, A., Wu, C., Marotta, S., Reposa, M., & Chong, S. (2019). Machine learning-enabled adaptation of information fusion software systems. *2019 22th International Conference on Information Fusion*, IEEE, 1–7.

[15] Pfeffer, A., Wu, C., Fry, G., Lu, K., Marotta, S., Reposa, M., Shi, Y., Kumar, T. S., Knoblock, C. A., & Parker, D. (2019). Software adaptation for an unmanned undersea vehicle. *IEEE Software*, *36*(2), 91–96.

SECURING AUTONOMOUS SYSTEMS – IT'S NOT JUST THE RADIO

Hal Aldridge, Ph.D.*

Communications security, be it simple point-to-point wireless or network, is still seen by many to be the only cybersecurity concern in uncrewed systems. While a significant threat, it is far from the only vulnerability an attacker could successfully use to target complex computer systems common in autonomous systems. IT and Defense industries have been addressing cyber threats for decades. These industries provide a wealth of knowledge in how attackers behave, how attacks are created, how security vulnerabilities are detected, and how security risks can be mitigated. However, a cybersecurity solution that works for an IT system may not be the ideal solution for an uncrewed system due to operational constraints, available computing capability, limited communications, reliability, and other factors. This paper will discuss evolving areas of cybersecurity significant to automated uncrewed systems and technologies/best practices used to address them.

INTRODUCTION

From Industrial to Defense applications, autonomous uncrewed systems have rapidly become part of essential operations. These systems will, in some cases, replace crewed systems eliminating the ability to revert to traditional operations. A cyber attack on a specific type of uncrewed system will likely not be targeted to one individual system, but to many different systems of the same type simultaneously enhancing its disruptive effect. The paradigm of single operator, single vehicle will also become much less common. Mostly found in less automated applications, in this paradigm the operator is connected to the uncrewed system by a dedicated communications link. Depending on the application, this can be a simple radio link to a complex satellite link. In this situation, a primary security concern is the security of the communication link. This communications security, be it simple point-to-point or network, is still seen by many to be the only cybersecurity concern in uncrewed systems.

Recently, with US Government restrictions on foreign made drones, supply chain has become a security concern in some applications but not others that are equally vulnerable but do not fall under the Government restrictions. For cybersecurity professionals, communication and supply chain security are significant threats, but they are far from the only vulnerabilities an attacker could successfully use to target complex computer systems common in autonomous systems. Developers and Operators of autonomous systems must not repeat the mistakes of Internet of Things and Industrial Control Systems applications by deploying unsecure capabilities and only adding security when forced by regulatory requirements or after a security incident.

IT and Defense industries have addressed cyber threats for decades where the volume and complexity of attacks only increase as technology matures and becomes more interconnected. These

* CEO, Secmation, 6601 Six Forks Road, Suite 470, Raleigh NC 27615.

industries provide a wealth of knowledge in how attackers behave, how attacks are created, how security vulnerabilities are detected, and how security risks can be mitigated. While this knowledge can be used in concept by autonomous systems, it must be properly implemented. A cybersecurity solution that works for an IT system may not be the ideal solution for an uncrewed system due to operational constraints, available computing capability, limited communications, reliability, and other factors. This paper discusses evolving areas of cybersecurity significant to automated uncrewed systems and the technologies/best practices used to address them. Topic areas addressed include proper use of cryptographic functions while maintaining performance, protection of customer data/intellectual property, software isolation, supply chain, and other key issues. Addressing evolving cybersecurity compliance requirements will also be discussed.

THINKING BEYOND THE RADIO/COMMUNICATIONS LINK

Like many new technologies, autonomous vehicles were initially developed without addressing cybersecurity. The development of new technologies is difficult and adding complications not relevant to solving core challenges, such as cybersecurity, is not seen as essential. As these technologies transition from research to initial applications, other considerations came into play. These can be safety, reliability, cost, performance, and other considerations beyond functionality. Cybersecurity initially became a consideration in uncrewed, teleoperated applications in the context of communications security (COMSEC).

In operational use, it was evident simple radio equipment could be used to monitor and possibly attack an unprotected wireless communications link. For monitoring, an observer using in some cases very low cost, readily available commercial receivers could monitor video transmissions from the vehicle to the operator. This monitoring could provide the observer access to sensitive data including indications of the location of the vehicle and its mission. Similar commercial equipment could be used to corrupt the telemetry link between vehicle and operator either leading to loss the operator's control of the vehicle or the ability of the attacker to take direct control of the vehicle.

These concerns led to initial requirements for COMSEC. While commonly referred to as encryption, COMSEC can use cryptographic based techniques for several objectives. These objectives are sometimes referred to as the CIA Triad: Confidentiality, Integrity, and Availability. Confidentiality protects information from disclosure to unauthorized parties. Integrity ensures that the data has not been modified from its original state. Integrity can also be used to identify the source of the data through "signing". Availability refers to the ability of authorized users to access the data when needed.

As the initial cybersecurity concerns were largely on confidentiality of video or other data being transmitted by the vehicle, the primary cryptographic method used to protect confidentiality, encryption, was the focus. As a side benefit, encryption can also provide some implied integrity protection but is not its intended use. Attacks on integrity can be indicated by corruption of encrypted data but there are well known attack methods (e.g. compromise of encryption key, replay attacks) that can violate the implied integrity. Cryptographic methods designed to provide integrity protection, some that can be integrated with encryption methods, provide much stronger integrity guarantees. Cryptography also has many different applications relevant to autonomous systems beyond confidentiality of the data link.

CRYPTOGRAPHY

Concerns

The need for encryption of communications links has many different solutions. These solutions can vary greatly in quality. Some designers chose to design their own solutions and, in some cases, encryption methods. This led to two key areas of concern: quality of the implementation and quality of the method.

On method, proven cryptographic methods/algorithms undergo rigorous review by mathematical, cryptographic, and other information security professionals. As a result, these methods and the assumptions they are based on are well vetted by the community. Any vulnerabilities identified during their use are normally quickly announced and addressed by the community. Assumptions regarding the period of time the algorithm will provide the required protections before it is "sunsets" are also provided and maintained. Evolutions in computation capabilities like quantum computing can enable an attacker to successfully break the cryptographic protections in a reasonable time period are a primary driver for sunset dates. However, little of this rigor is commonly seen in non-standard algorithms. While non-standard algorithms can be seen as novel, providing some uniqueness an attacker must figure out or be designed to address system limitations such as computation or communications bandwidth, it is usually the case these non-standard solutions are not as robust as proven solutions.

Regarding quality, this has been an ongoing area of confusion within multiple industries. The key point is that most successful attacks on proven cryptographic methods do not succeed by targeting vulnerabilities of the method itself, but by targeting vulnerabilities in the implementation of the method. These implementation vulnerabilities can be caused by coding errors, incomplete testing, violating usage assumptions, or other causes unrelated to the method itself. As result, the best practice is to use cryptographic methods from proven sources/vendors that passed independent testing. A common open-source cryptographic library widely used is OpenSSL. While OpenSSL is used in applications from IT to consumer devices to space, it shows that even a widely used, well reviewed library can have vulnerabilities such as the well documented Heartbleed. Another item affecting quality is the strength of the algorithm itself. Strength can be related to algorithm selected, mode selected, and encryption key length among other factors. These parameters are often represented in algorithm shorthand such as AES-256-GCM to identify the specifics.

Cryptographic Certifications

A key cause of confusion on quality is specific requirements on certification of the cryptographic implementation. In general, the goal of a certification for any function (e.g. safety, interoperability) is to show a specific implementation is implemented with compliance to a standard through testing by a recognized independent third party with the knowledge and qualifications needed to perform the testing.

A common standard for certification of cryptographic modules is FIPS 140-2 and its replacement FIPS 140-3. These standards are maintained by NIST. Cryptographic modules can consist of software only implementations of cryptographic algorithms or a combination of hardware and software which implements the algorithms. While the details of the FIPS 140-2 and 140-3 standards are beyond the scope of this discussion, two key concepts are relevant. First, a FIPS cryptographic module certification requires that the implementation has been tested by an independent, NIST approved laboratory. This process ensures the implementation was adequately reviewed and tested. After testing, NIST reviews the results and certifies the module and a list of approved modules can be found on the NIST website. Second, the compliance with the FIPS 140-2 standard (certifications valid for 5 years after validation or through September 21, 2026) and the newer FIPS 140-3 standard are common in requirements for Industrial and Government security requirements for procurement of IT solutions, communications, and computer equipment. These requirements are becoming more common for autonomous and uncrewed vehicle procurements.

Specific to certain US Government and Defense applications, there can be a requirement for COMSEC to protect classified information. There are similar requirements for these applications in other countries. In the US, an equivalent role to NIST, but for classified COMSEC, is filled by NSA. In general, FIPS 140-2 or 140-3 certifications are not considered adequate to protect classified information. As a result, NSA provides guidance and approves certifications for classified data protection using a different process. The process can be adapted for different levels of classification.

Especially in Government and Defense applications, overlaps between similar organizations referencing different requirements for similar equipment is a source of confusion for vendors. Vendors should consult their customers and information security professionals to understand the reason for the requirements and methods to comply with them. For many situations, there are ways to take existing solutions and integrate/adapt them to autonomous vehicle applications to meet many different customer requirements.

Beyond Encryption

While encryption is common in COMSEC, it has its role and its disadvantages. While encryption is the preferred method for confidentiality protection, it can require significant computational resources, introduce latency in timing, increase communications bandwidth, decrease communications reliability, and other undesirable effects for autonomous systems. Reducing the quality of encryption, increasing computational or communications resources, and/or not using encryption are common workarounds for these undesirable affects. Good cybersecurity system design is a preferable solution to address both cost and performance issues of encryption.

A consideration in cybersecurity system design is what data must have confidentiality protection. It should be noted, in many cases not all data in the communications link require confidentiality protection. It is commonly possible to separate confidential and non-confidential information in separate packets, logical links, or other methods reducing the amount of data to be encrypted (and decrypted). This reduction has a direct effect on system resource requirements.

In some cases, there is critical data that does not have confidentiality requirements but does have integrity requirements. An example could be vehicle status information relayed from the vehicle to the operator. In some applications this data could be considered non-confidential. However, attacks on the integrity of this data could result in the operator or autonomous system making improper decisions on bad data. In general, cryptographic methods such as HMAC can provide integrity authentication protection at significantly lower resource cost and latency than encryption. Similar in purpose to a checksum, the cryptographic methods provide additional assurances the data initially transmitted was not modified when received.

Beyond COMSEC

Communications is not the only use of information in an autonomous system. These systems store information as software code, configuration parameters, mission data, and other items needed to perform their function or collected as part of performance of the function. In many applications, loss or corruption of this information has not been considered to be a risk and should be reconsidered in many cases. Threats to autonomous vehicles are changing. These systems may contain sensitive data including intellectual property (e.g. software, trade secret methods, machine learning algorithms), customer proprietary data, and data covered by Government privacy regulations. As systems become more autonomous and numerous, the risk of physical loss of systems through theft or other actions will grow. Attackers with physical control of systems have proven methods they can use to extract information stored. Autonomous systems will be increasingly deployed with connection to larger networks including the internet. As a result, they will be vulnerable to the same type of "hacking" attacks that have proliferated in Internet of Things and Industrial Control Systems applications. Successful attacks can provide a foothold in the system that can be used to exfiltrate unprotected information in the system in some cases using the existing COMSEC protected communication link.

Data stored in the system is referred to as data at rest. This differs from data in motion referring to data on the communication link and in active processing computer systems. Data at rest can be confidentiality and/or integrity protected similarly to data in motion on the communications link. A common example of data at rest encryption is full hard disk encryption such as Microsoft's BitLocker. Like COMSEC, data at rest encryption has similar undesirable effects. These can be more significant than in COMSEC due to increases in data volume and effects of latency on internal processing. Also similarly, good cybersecurity systems design can be used to mitigate these effects in similar manners to COMSEC.

A specific case of data at rest protection is information used in the process to boot the system. Most security architectures rely on the assumption that the system starts in a secure state and the concern during operation is maintenance of that secure state. An attack on the data used in the boot process to obtain the initial secure state can make all other system security ineffective and can mask the real insecure state of the system. A secure boot process that verifies the integrity of the boot information (e.g. software, parameters) and, if required, decrypts the boot information from its protected at rest state can provide this initial secure state. Many modern processors provide the hardware support needed to implement a secure boot process. The quality of this process and effect of boot-up time are considerations on processor selection and boot process design. An important point is while many systems are capable of implementing a secure boot process, they do not actually implement one. This practice should be reconsidered moving forward or any additional investment in system security can be at risk.

SOFTWARE ISOLATION

The evolution of autonomous systems is changing the assumptions on software quality and security. Early systems commonly used custom designed software written and tested by the system developer. As a result, the pedigree of this software was fully in control of the developer. As autonomous systems have evolved, they have begun to make use of open-source software libraries, third-party software, and open architectures similarly to many other computing applications. As a result, the developer does not have full knowledge of the software executing on their systems. This raises concerns for reliability, safety, and security. As a practical matter, in most systems this situation cannot be fully avoided, but there are techniques that can be used to mitigate the associated risks. Many of these techniques are related to security concepts of isolation and least privilege.

Software and data isolation involve separating the execution of software and controlling access to data and system resources. The concept of least privilege is where software applications are only granted access to the data and system resources they need for proper operation. While this may sound obvious, in practice many systems are designed the opposite way. In these systems, applications have access to all data and system resources unless specifically limited. This provides maximum flexibility and ease of programming, but can put a system at risk to either a malicious application or simple errors in one application that can have significant effect on system operation unrelated to the application's purpose. Consider a gaming application that corrupts sensitive accounting data because it was not properly isolated from the accounting application's data.

Software execution can be isolated through proper system design using proven techniques. The most common is the use of an operating system's process separation and privileges. Processes can be configured by the designer to have specific access. However, this is not always used properly and/or the least privilege configuration maintained through software updates and changes in design over the product lifecycle.

A concept first used in IT is to group related processes, data, and applications together and isolate them as a group. Different implementations of this concept include virtual machines and containers. These different implementations each have their own strengths and weaknesses. A key weakness can be the additional processing power and system resources including memory required to implement them. Until recently, the embedded computing hardware used in autonomous vehicle applications was taxed to implement the required functions and did not have the capability to support these more advanced and capable isolation mechanisms. Modern embedded systems provide additional design options including processors with hardware acceleration of key functions needed for these isolation mechanisms and more system resources enabling their more common use. Autonomous systems designs should consider the use of these isolation mechanisms for benefits in cybersecurity, safety, and maintainability.

SUPPLY CHAIN

Hardware

The security of the hardware supply chain for components of autonomous systems has been highlighted over the past few years. In one sense security is about part availability of critical components needed to build computing system incorporated into autonomous systems. In another sense security of the supply chain is specific to the cybersecurity of components. Third party components that incorporate software, firmware, and computing hardware can be corrupted with malicious code intentionally by the designer, as part of a security breach in the development or manufacturing process, or as part of an update during the lifecycle of the component. Any of these pathways can lead to a compromise in the autonomous system they are incorporated into.

Aside from malicious code in the hardware there are two specific supply chain concerns related hardware: counterfeiting and source of origin. Counterfeiting of electronic components is an ongoing issue across industries. Components that appear genuine may not function properly, may not meet functional specifications of genuine systems, or may be designed to maliciously compromise a system. Some malicious counterfeits can perform exactly as genuine components while also performing a malicious activity such as exfiltrating information.

Source of origin is a second concern. This concern has directly affected the autonomous vehicle industry through the banning of uncrewed systems and related components for certain origins by the US Department of Defense and multiple Government agencies. This ban is the product of security analysis of systems and components which concluded these components and systems performed actions that resulted in a significant cybersecurity risk in many critical applications. As a result, limitation on the procurement and use for complete systems and key components used to build these systems have reduced the options for operators and developers of autonomous systems decreasing availability and increasing costs. While these bans have not been adopted by the wider market, this is an example of how significant supply chain cybersecurity issues can affect specific segments and applications within an industry.

Software

Until recently, supply chain was seen to be primarily a hardware issue. It is now becoming a concern in software. As open source and third party commercial software makes up larger proportions of most software implementations, it has become difficult for a developer or an operator to know what software is contained in their systems. A cybersecurity example why this is a concern is the OpenSSL Heartbleed vulnerability mentioned previously. When discovered, it was obvious this was a very significant security vulnerability that must be addressed. The issue was determining what systems were affected. OpenSSL is used in millions of systems worldwide and the vulnerability was also related to specific versions of OpenSSL. Developers and OEMs of systems in many cases were unsure which version of OpenSSL was deployed in which of their systems. Operators of systems could not determine if their systems used OpenSSL and/or the version they used. This situation is especially problematic in legacy systems that may no longer be supported by the manufacturer leaving the operator with no direct way to confirm the status of their system.

A solution to this problem that is rapidly being adopted is the Software Bill of Materials (SBOM). A SBOM provides a tracible listing of software components and versions incorporated into a system. This gives a system developer a way to track what software they are incorporating into their solution. The solution developer/OEM would then provide an integrated SBOM for their system or subsystem to the end user/operator. Automated tools can be used to track vulnerability reports on software components and flag systems containing the components for remediation. While simple in concept, the process of creating, maintaining, and obtaining SBOM is difficult. Many vendors are not ready to comply with purchasing requirements to provide SBOMs and operators are not familiar with their applicability. Autonomous systems and components (both hardware and software) should prepare to utilize SBOMs for their systems and be able to provide SBOMs to their customers.

EVOLVING CYBERSECURITY CERTIFICATION REQUIREMENTS

Referenced earlier in the discussion on Cryptographic Certifications, cybersecurity related certifications are becoming more common in the autonomous systems industry. As this paper has discussed, there are many cybersecurity concerns in autonomous systems beyond COMSEC and cryptography. The same need for standards and independent testing for these broader cybersecurity requirements is necessary for the same reasons as cryptography. Like cryptography, the maturity of the industry and understanding the real affects of cybersecurity attacks in safety, cost, and availability is driving the need for end users/operators to understand the quality of the cybersecurity protections in a system.

An industry example of cybersecurity certifications are BlueUAS and the newer GreenUAS. BlueUAS was developed by the US Defense Innovation Unit (DIU) in response to the need for commercial small UAS to comply with restrictions related to supply chain that led to banning of many existing UAS because of their source of origin. The BlueUAS standard focused on basic cybersecurity protections in addition to COMSEC, along with restrictions on source of origin of systems and components. UAS OEMs can be certified to this standard through a process managed by DIU and placed onto an approved list. By writing BlueUAS requirements into their procurements, Defense and other users can purchase systems certified to be in compliance with regulations on supply chain and contain independently verified cybersecurity protections adequate for many applications of small UAS.

The Association of Uncrewed Vehicle Systems (AUVSI) has recently announced the GreenUAS certification. Developed in cooperation with DIU, GreenUAS incorporates the cybersecurity and supply chain requirements of BlueUAS and extends them with some selected cybersecurity best practices to make the standard more capable without adding significant complexity over BlueUAS. The certification process for GreenUAS is managed by AUVSI. AUVSI management promises to enable OEMs to have their systems certified quicker and thereby expand the availability of secure systems. As an industry driven standard, GreenUAS promises to enable a broader community of end users/operators to adopt the standard in their procurements leading to a larger market for secure small UAS.

While BlueUAS and GreenUAS are positive steps in the evolution of the autonomous systems industry to adopt cybersecurity, they are only a first step. A key concern is that the cybersecurity requirements in these standards, while a significant improvement in those previously used on small UAS, are limited. Compared to the cybersecurity requirements commonly applied to and certified in aerospace and IT applications, BlueUAS/GreenUAS contain a small subset which can lead to systems vulnerable to common attack types.

Like safety, cybersecurity is really concerned with risk mitigation. The BlueUAS and GreenUAS certification programs are designed specifically to address risks common in small UAS adequately for many applications. For more critical applications and/or larger, higher value autonomous systems, additional cybersecurity protections must be considered to address the risks associated with these systems. These standards are currently in work by many Government and Industry groups. As a result, autonomous systems developers should be planning on how to address additional cybersecurity requirements and support certification processes for both their new and current systems to remain viable in key markets.

CONCLUSION

This paper has discussed the multiple cybersecurity concerns for autonomous systems "beyond the radio". These concerns are related to cryptography including its different applications, software isolation in complex systems with components from multiple sources, and both hardware software supply chain concerns. This is not a comprehensive list of cybersecurity concerns in autonomous systems. However, addressing these concerns is a necessary part of the path to more secure autonomous systems that will have the trust of their users. Cybersecurity standards and certifications are evolving to provide assurance in the quality of cybersecurity implementations specific to the risks and needs applicable to autonomous systems. Developers, OEMs, and end users/operators should prepare for new cybersecurity requirements on autonomous systems which will enable these systems to achieve their potential.

VNX+: A MODULAR, SCALABLE COMPUTING ARCHITECTURE FOR AUTONOMOUS VEHICLES

Mark Littlefield*

While still a work in progress within the VME International Trade Association (VITA), VITA 90 VNX+ is maturing rapidly. This rugged, modular small form factor module hardware architecture is an attractive approach for autonomous vehicles regardless of their deployment environment (air, ground, or sea). VITA 90 offers significant enhancements above the earlier VITA 74 VNX standard, including more tightly defined Utility Segment, Communications Segment, and new Overlay Segment, provisions for blind-mate optical and coax connections, power supply module definitions, and well-defined slot profiles for payload modules, switches, PNT/Radial Clock modules, and even space payloads. Elements of VITA 90 have also been adopted by the The Open Group® Sensor Open System Architecture™ (SOSA™) to address their needs for a small form factor "significantly smaller than 3U VPX", but still in alignment with SOSA principles of modularity, scalability, interoperability, and replaceability. In this paper we will review the current status of the various VITA 90.x standards documents, as well as details of the adoption of VNX+ by SOSA. Finally, we will examine how elements of VNX+ can be applied to autonomous vehicles requiring small form factor computing hardware.

INTRODUCTION

In December 2017 ANSI approved the VITA 74/VNX Small Form Factor Module Base Standard[1]. This document defined a pair of standardized plug-in modules intended to address what was seen at the time as a growing need for backplane-based plug-in cards smaller than other formats such as 3U VPX as defined by VITA 46[2] and VITA 65[3] but providing much of the same functionality as these other formats including fast serial fabric interconnects and naturally rugged construction.

VITA 74 defined two module formats, each 89mm by 78mm but differing in their width (12.5mm and 19mm respectively) (Figure 1). The 12.5mm module features a 200 pin connector, while the wider 19mm module features a 400 pin, and both feature alignment features to aid in module insertion into a backplane. The nominal use case for VNX modules includes a backplane with two or more slots, a mechanical frame to hold the backplane and VNX modules, a chassis front-panel to provide I/O for the system, and cooling plates attached to the chassis frame and making contact to the individual VNX modules to provide a conduction cooling path (Figure 2). The standard is silent with respect to defining the cooling structure, but commonly implementations used passive fins or cooling channels fed by a fan.

* Director, Systems Products, Elma Electronic, 44350 S Grimmer Blvd, Fremont, CA 94538, Mark.Littlefield@Elma.com

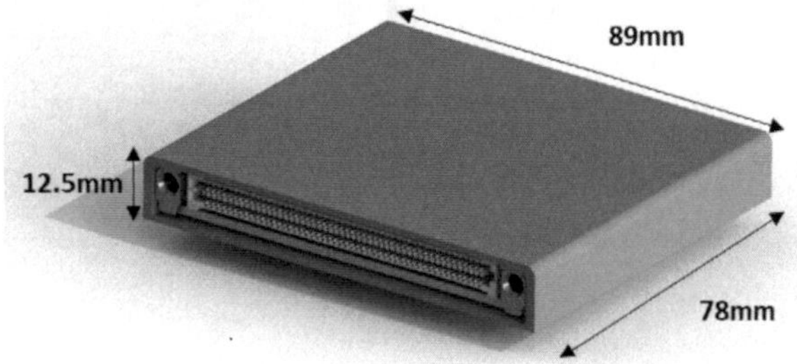

Figure 2. VITA 90 (VNX+) 12.5 and 19mm modules. From VITA 90.0 (draft)

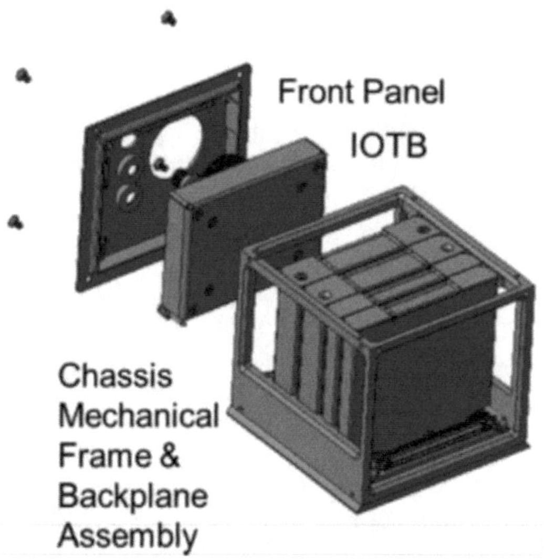

Figure 1. Example of a typical VNX assembly. From VITA 74.0.

The standard also defined a pin map consisting of a Utility Segment S0 (power and ground pins, I2C System Management port, Reset, Geographical Addresses, and a common System Clock), a Serial Fabric S1 region consisting of PCI Express and user defined serial fabric pairs, and a User Defined S2 region with single-ended and differential pairs.

VNX gained some modest initial success, but generally failed to generate a wide market acceptance enjoyed by other backplane-based form factors such as OpenVPX. However, with its adoption into the Sensor Open Systems Architecture[4] SOSA™ Technical Standard[5] and several SOSA identified enhancements added (described below), VITA decided to form a new VITA 90 standard.

VITA 90 is broken into several sub-documents:

- VITA 90.0: Baseline standard
- VITA 90.1: Slot Profiles
- VITA 90.2: Aperture Fills

- VITA 90.3: Power Supplies
- VITA 90.4: Cooling and Mounting Systems
- VITA 90.5: Space VNX+

While at the time of writing all of these documents are still in draft form, several are mature enough that hardware suppliers are beginning new product designs based on the draft standards, and several integration projects have started expecting to use these products. In addition, the SOSA Technical Standard has adopted the 19mm module format, several of the slot profiles defined in VITA 90.1, several aperture fill solutions defined in VITA 90.2, and several other key components of the VITA 90.3/4/5 documents as part of the SOSA Reference Architecture.

The rest of this paper is devoted to:

- Describing the nature of the enhancements identified as needed by the SOSA Small form Factor Subcommittee (SFFSC)
- Describing the important technical details presently captured in the VITA 90.x documents
- Describing what has been captured in the latest SOSA Technical Standard (Version 2.0 Snapshot 1) and why
- Examining how VNX+ can lend itself to uncrewed and autonomous systems, be they air, land, or sea-based

THE SENSOR OPEN SYSTEMS ARCHITECTURE (SOSA), SMALL FORM FACTOR SUBCOMMITTEE, AND VNX+

The Sensor Open Systems Architecture, or SOSA, is a technical standard written and maintained by the The Open Group® SOSA™ Consortium, a group consisting of government, industry, and academia representatives formed to define a reference hardware and software architecture for advanced command, control, communications, computers, cyber, intelligence, surveillance, and reconnaissance (C5ISR) systems. Specific applications for SOSA include (but aren't limited to) radar, signals intelligence (SIGINT), electronic warfare (EW), electro-optic/infrared (EO/IR), and communications systems. SOSA achieves this by leveraging existing standards, limiting options provided by these standards to a minimum practical set, and creating new standards language when needed.

SOSA has broad adoption across all three of the US services, as well as the US Intelligence community. This broad adoption also means that industry is responding with an ever-growing array of products based on the standard.

SOSA is based on a set of guiding principles, including:

- Modular
- Interoperable
- Replaceable
- Securable
- Portable
- Plug-and-Play
- Scalable
- Resilient
- Vendor Agnostic
- Royalty Free
- Protection of IP
- Extensible and Evolvable
- Simple through complex systems
- Small through large systems

As a result, any hardware standard adopted by the standard must adhere to these principles.

In late 2020 the SOSA™ (Sensor Open Systems Architecture) Consortium stood up a new Small Form Factors subcommittee (SFFSC). This group had the charter to "explore options for hardware 'physically smaller than 3U VPX' and to draft standards language for inclusion into the SOSA

Technical Standard." Among the first tasks of the subcommittee was to perform a trade study to explore existing small form factor options. This study revealed two potential candidates for SOSA SFF:

- "Short" VPX: a new variant of 3U VPX that is 100mm deep rather than the typical 160mm, but otherwise the same. However, this only resulted in a roughly 30% reduction in size on a per-slot basis, and did not address some important use cases.
- VNX: Potentially very attractive and offering a roughly 70% reduction in per-slot size. However, VITA 74 was found to be lacking several features important to alignment with SOSA principles.

The SFFSC decided to adopt both, and to take on the task to drive enhancements to VNX to address its shortcomings and to bring it into alignment with SOSA principles. This work later shifted to a new VITA 90 (VNX+) working group as the VITA leadership decided that the changes introduced for SOSA alignment were substantive enough to warrant a new standard being drafted.

SHORTCOMINGS OF VITA 74/VNX IDENTIFIED BY SOSA

As mentioned in the introduction, the SOSA SFFSC identified a number of shortcomings in the original VITA 74/VNX standard that would need to be addressed in order to make VNX align with the SOSA principles described above and to make it a suitable form factor for inclusion in the SOSA Technical Standard. These shortcomings can be generally grouped into five categories:

- There were no provisions for blind-mate coax or optical connectors (necessary for SOSA sensor processing applications)
- The utility segment was not well aligned with SOSA
- Specific slot profiles would need to be created (with no user-defined pins)
- A better understanding of plug-in card thermal characteristics was needed
- Mechanical enhancements for modularity and replacability were needed

Preliminary work on each of these was done within the SOSA SFFSC during 2021. In late 2021 it was decided that this work would be better captured in a new VITA 90 standard and simply referenced in the SOSA technical standard, so the material was transferred to VITA. Initial draft documents were created, and work continued to flesh them out through 2022 and early 2023.

VITA 90.2: Blind Mate Connectors

A critical feature of OpenVPX is the availability of blind mate coax and optical connections (as defined in the various VITA 66.x and 67.x documents[6 7 8]). Blind-mate connectors allow one to avoid the reliability issues and general "messiness" of front-panel connectors by mating the connectors by simply inserting a card into its slot. In addition, conduction-cooled OpenVPX boards generally try to avoid front-panel connectors, so blind mate connectors are the preferred connectivity method in these cases.

For VNX+ blind mate connectors are especially critical as VNX+ has no concept of front-panel I/O. In addition, the "front panel" of a VNX+ plug-in card is usually part of the heat transfer path for the card, further discouraging any sort of front-panel I/O mechanism. Fortunately, Samtec provides their SeaRay connector in shorter versions than the 400-pin version called out in VITA 74.

VITA 90.2 adopted the 240- and 320-pin SeaRay connectors in addition to the original 400-pin. This provides either a full or half-aperture for coax or optical connections. To fill this space, a set of aperture fill blocks have been defined and captured in VITA 90.2. Figure 3 shows the four aperture fill blocks currently captured by the standard.

Connector/Aperture Combination	Config Class	Connector Type	Aperture Fill Config	Image
VNX+ Connector Only	1	400-pin	None	
VNX+ w/ Small Aperture	2	320-pin	2x MT	
VNX+ w/ Small Aperture	3	320-pin	1x MT 6x RF	
VNX+ w/ Large Aperture	4	240-pin	2x MT 20x RF	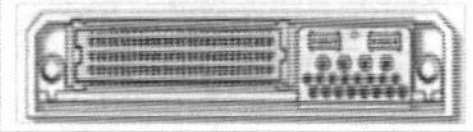

Figure 3. VNX+ module connectors approved by SOSA. From The SOSA Technical Standard

S0 Utility Segment and S1 Communications Segment Organization

The original VITA 74 S0 Utility Segment was not well aligned with the needs of SOSA. In addition, there were numerous pins reserved for future use (RFU) in the pin map that SOSA sought to make use of. Therefore, a redesign of the S0 was needed. This also gave the team an opportunity to create a payload profile "template" with a common S0 Utility Segment, a common S1 Communications Segment (where the major board-to-board communications planes would be located), and a new S2 Overlay Segment consisting of a region of copper pins and an aperture region (depending on the connector) that would be used for specialized slot profiles. Lastly, in the years since VITA 74 was first ratified, it was learned that pins in the "lower" rows have noticeably better signal integrity than the "upper" rows. Thus, there was a desire to "push" low-speed signals and power to the upper rows, leaving the lower for higher-speed communications.

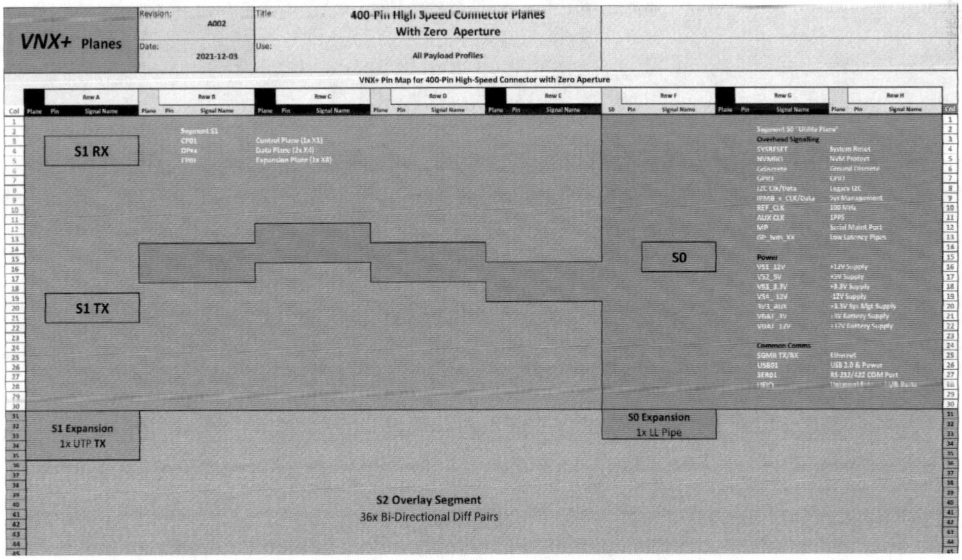

Figure 4. VITA 90 400-pin defined regions. S0 Utility (green), S1 Communications (Orange), and S2 Overlay (lavender) Segments.

This reorganization resulted in the region map shown in Figure 4. In this figure the pin map is rotated 90 degrees so that the "lower" pins are on left and the "upper" pins are on the right. The S0 Utility Segment pins are marked in green, the S1 Communications Segment pins are marked in orange, and the S2 Overlay Segment region is marked in lavender. Note, too, that the S1 Communications Segment transmit and receive pairs are physically separated to further improve overall signal integrity.

Diving deeper into the S0 Utility Segment there are several regions of interest (see Figure 5). First are the power planes and utility signals common to OpenVPX, including NVMRO (Non-Volatile Memory Read Only), GDiscrete1, SYSRESET, the Geographical Addressing pins, and the dual IPMB management busses. Unlike OpenVPX, however, the standard adds several communications ports common to many plug in cards, including a serial Maintenance Port, a user TIA-232 port, USB, GPIOs, and a Gigabit Ethernet port (MDI only – the magnetics live off-board on an I/O interface board).

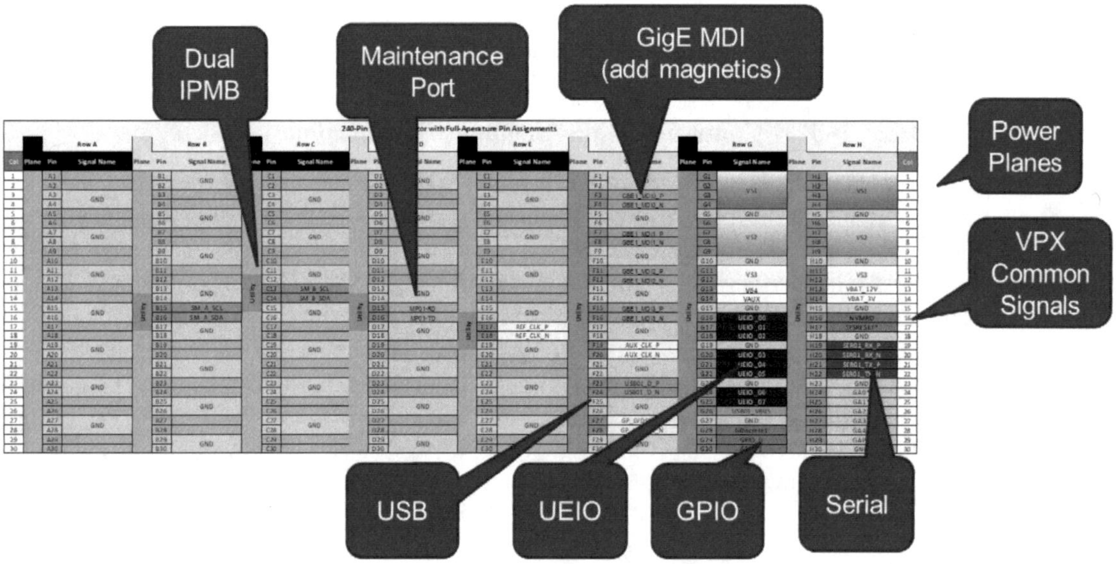

Figure 5. S0 Utility Segment features

Lastly, a new region called Unique External I/O was added. VNX+ specifically targets small integrated systems, and these systems often have need for additional custom I/O, electo-mechanical devices like relays and servos, and so on. Thus, there was a desire to have a region of the S0 Utility Segment defined to interface with such devices.

Multiple options for UEIO have been defined in VITA 90, as shown in Figure 6. As SOSA principles drive towards common implementations, it is expected that only one of these options will be supported in the SOSA Technical Standard Version 2.0.

- **1x to 4x I2C**
- **1x to 2x SPI**
- **2x I2C and 1x SPI**
- **1x I2C and 1x SPI with 1x-3x SPI Selects**
- **1x SPI with 1x-5x SPI Selects**
- **2x SPI with 1x SPI Select each**

Pin	Utility
G14	VAUX
G15	GND
G16	UEIO _00
G17	UEIO _01
G18	UEIO _02
G19	GND
G20	UEIO _03
G21	UEIO _04
G22	UEIO _05
G23	GND
G24	UEIO _06
G25	UEIO _07
G26	USB01_VBUS
G27	GND

Figure 6. VITA 90 (VMX+) Unique External I/O options

As mentioned previously, the S1 Communications Segment is where the main intra-system communications ports are all located. OpenVPX and SOSA both rely heavily on the concept separate communications planes, including a Data Plane (where major board-to-board communications take place), a Control Plane (a separate communications channel where command messages are passed), and an Expansion Plane (where tight board-to-board communications links like CPU-to-GPU using PCI Express or FPGA-to-FPGA using Aurora are established). VNX+ defines two four-lane Data Plane ports, one or two Control Plane ports, and eight lanes of Expansion Plane, all at speeds up to 25Gbps[9]. Details vary slightly between the three connector sizes (240, 320, and 400), and can be found below in the slot profile descriptions.

SOSA VNX+ SLOT PROFILES

Multiple slot profiles have been defined in VITA 90.1 (VNX+ Slot Profiles) and VITA 90.5 (Space VNX+). SOSA has chosen a subset of these for inclusion into the technical standard. At the time of publishing of this document the SOSA Technical Standard Version 2 Snapshot 1 calls out the following profiles:

- 240 Pin Payload
- 400 Pin Switch
  - 7x Data Plane, 8x Control Plane
- 320 Pin Switch
  - 5x Data Plane, 5x Control Plane
  - Aperture for two MTs and up to 48 ferrules
- 7-Slot Radial Clock
  - Based on the Payload Profile Template
  - 7x REF_CLK/AUX_CLK pairs
- High-density 320 pin Radial Clock
  - Not based on the template
  - 22x REF_CLK/AUX_CLK pairs

Slot profiles for 320 and 400 pin payloads, as well as space-specific slot profiles are expected to be defined in the upcoming versions of the SOSA Technical Standard.

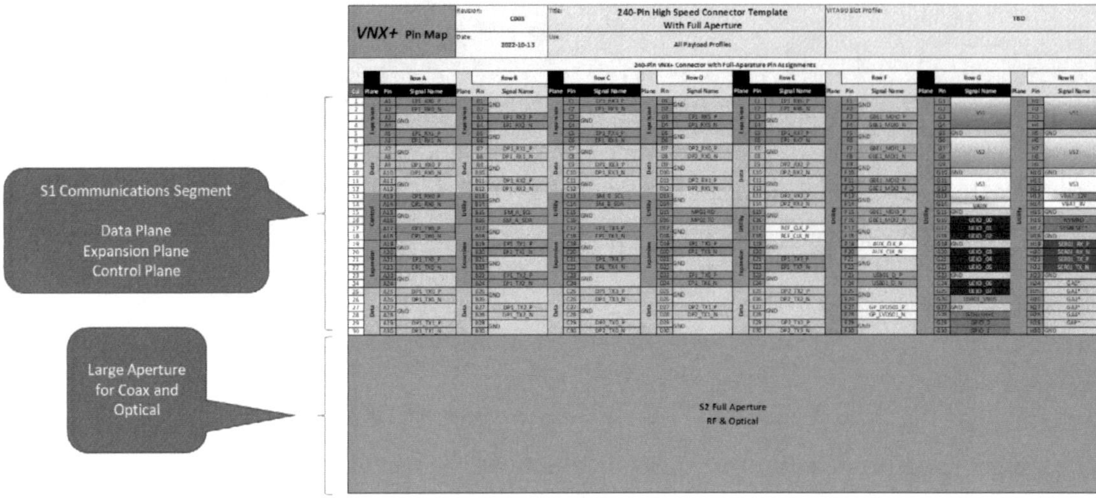

Figure 7. 240-pin Payload Profile Template with the S0 and S1 regions marked. Note that in this case it also forms the natural 240-pin Payload Profile

240-PIN PAYLOAD PROFILE

The 240-Pin Payload profile selected by SOSA is shown in Figure 7. It is, quite simply, the Payload Profile Template mentioned above as applied to the 240 pin connector. The S1 Communications Segment has two Data Planes, one Control Plane, and eight lanes of Expansion Plane. The S2 Overlay Segment is simply the large aperture, supporting the aperture fill block defined above.

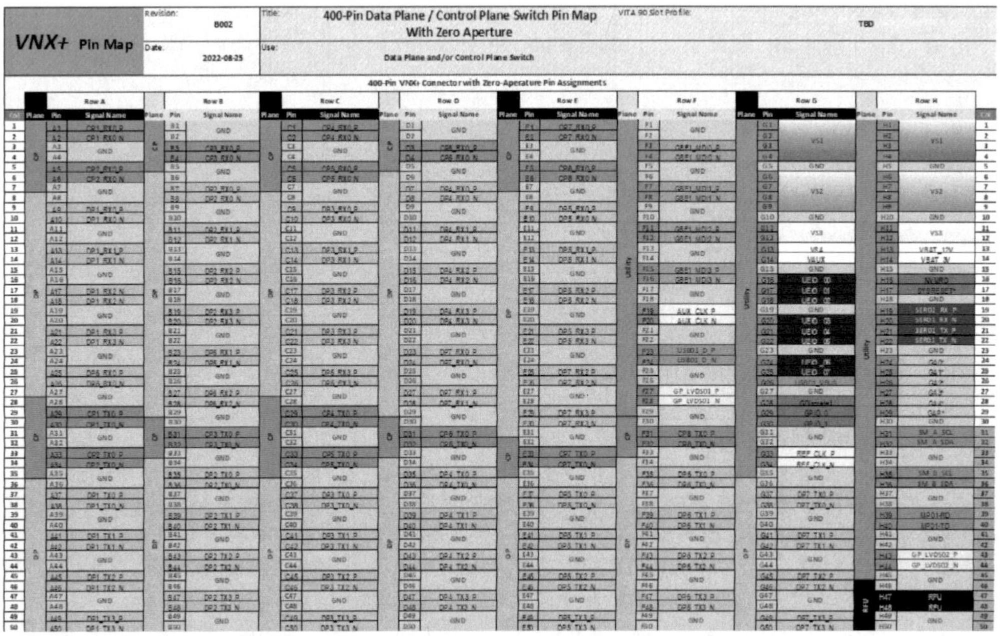

Figure 8. 400-pin Switch Profile

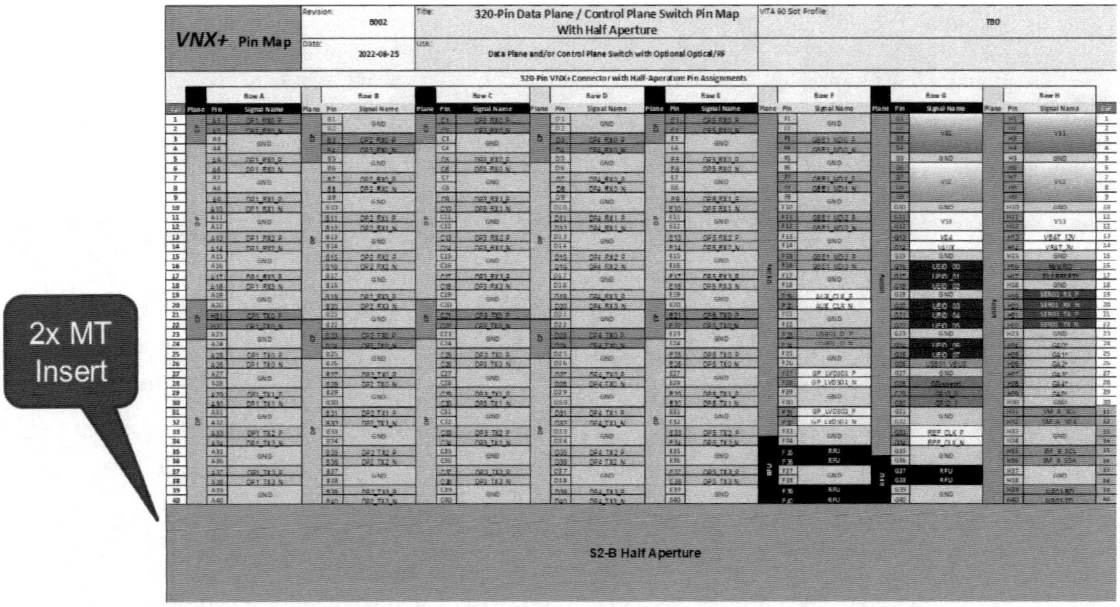

Figure 9. 320-pin Switch Profile with the aperture for the two-MT aperture fill block marked

SOSA DEFINED SWITCH PROFILES

Two switch profiles were chosen by SOSA, one using the 400 pin connector, and one using the 320 pin connector and supporting the dual MT aperture fill block. The 400-pin version supports up to seven Data Plane and eight Control plane ports, while the 320-pin version supports five Data Plane and five Control Plane ports. Figures 8 and 9 show these two profiles.

SOSA DEFINED RADIAL CLOCK PROFILES

Sensor systems like those targeted by SOSA often use radial clocks to provide a very precise time synchronization between payload cards. REF_CLK is defined in SOSA as a high-precision 10 MHz signal, while AUX_CLK is defined as a 1 PPS signal. Both of these signals are usually derived from a high-quality GPS source.

SOSA decided to pick two radial clock profiles for the technical standard, both based on the 320-pin connector and using the hybrid aperture fill block to provide coaxial connections to a central clock source. This decision was driven by two competing requirements. First, there was a desire to leverage the Payload Profile Template and to place the REF_CLK and AUX_CLK pairs in the S1 Overlay Segment, which resulted in a board that could support up to seven payload cards, while still acting as a payload itself, such as a single-board computer (SBC).

However, there was a desire to support much larger VNX+ systems. While this may seem contradictory for a "small form factor" platform, there are clearly applications that require a very compact modular form factor but which also needs to scale to a relatively large number of plug-in cards – imagine a common launch tube application or a missile-shaped instrumentation pod. In this case, a profile supporting more than seven payload cards was needed.

Figure 10 shows the two SOSA defined radial clock profiles.

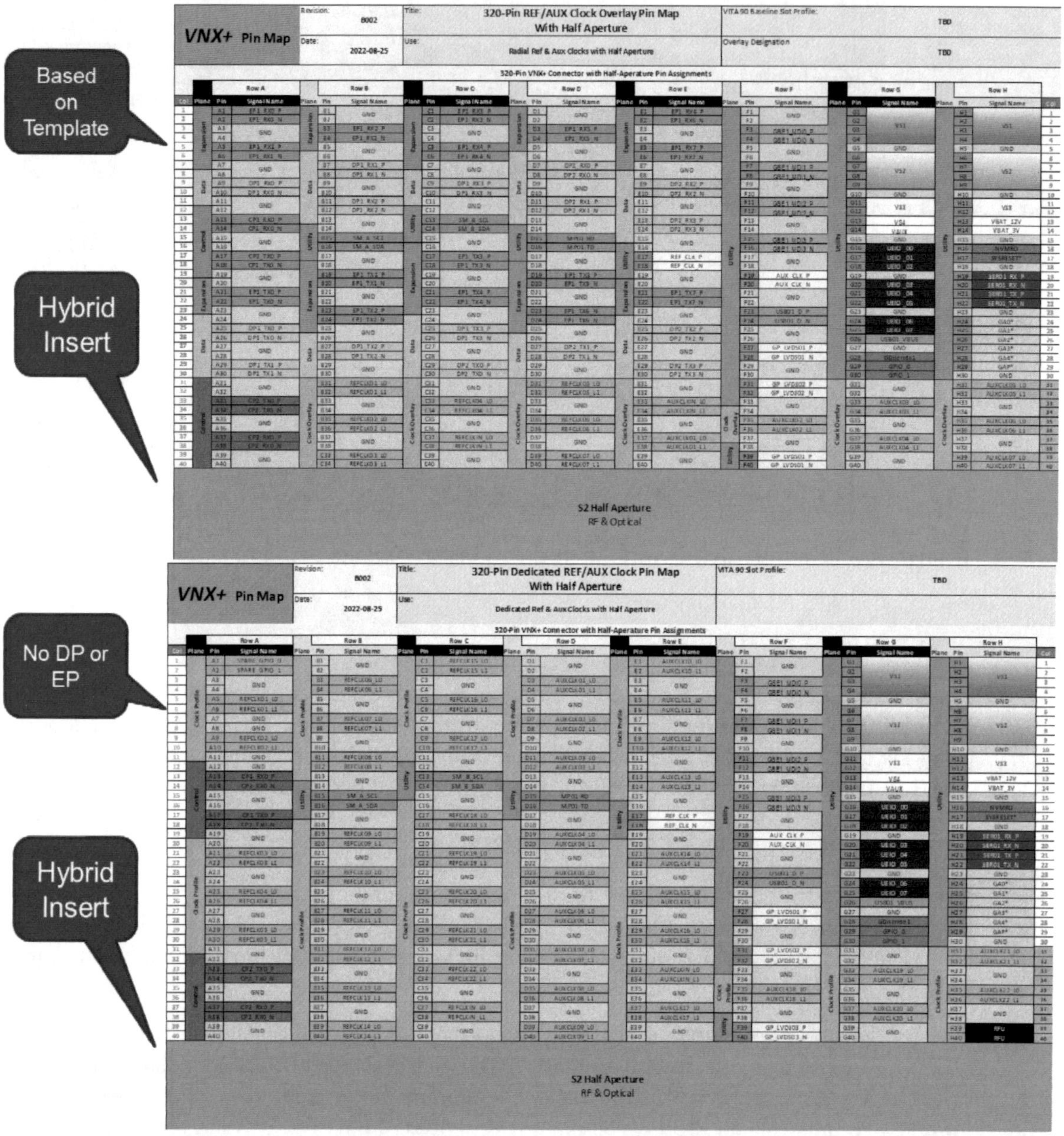

Figure 10. Two Radial Clock Profiles adopted by SOSA. The upper is based on the Payload Profile Template and supports seven REF_CLK and AUX_CLK pairs. The lower is unique and supports up to 22 REF_CLK and AUX_CLK pairs.

320- AND 400-PIN PAYLOADS AND SINGLE-BOARD COMPUTERS

While SOSA has not yet defined Payload slot profiles using the 320- and 400-pin VNX+ connectors, VITA has defined several. In this section we will examine one of these slot profiles. We can expect that this profile is probably fairly representative of what SOSA will ultimately define for 320- and 400-pin Payload profiles.

Figure 11 shows one of the several payload slot profiles captured in the draft VITA 90.0 document. This profile, like the others, is based on the Payload Profile Template described previously. In the S2 Overlay Segment we find several features defined:

- A USB3 port, with power
- A video port, suitable for Display Port or HDMI. Note that SOSA already defines Display Port 1.2 as a supported protocol
- A new Peripheral Port – a 2-lane port expected to be used for closely-attached peripherals such as SATA or NVMe storage, ARINC 818 video input, etc.
- 42 Mezzanine Mapped pins – for differential or single-ended I/O from an on-board mezzanine card

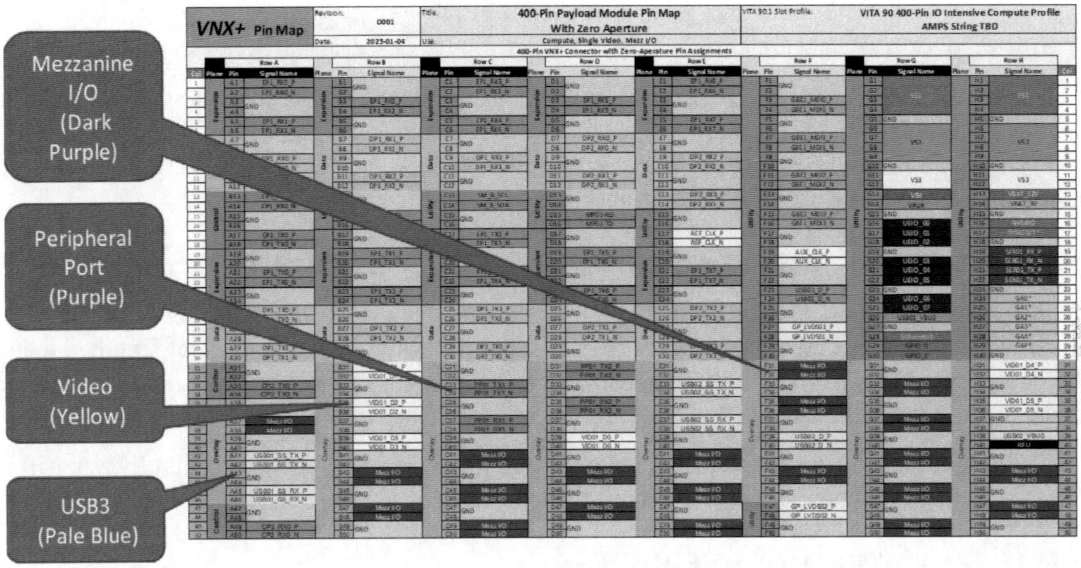

Figure 11.Example 400-pin Payload Profile. While this is captured in the VITA 90 document, no 400-pin profile has been defined yet in the SOSA technical standard, so this should be treated as an example only.

At the moment there is no industry-standard mezzanine small enough to fit inside of a VNX+ plug-in card that provides a convenient mapped set of I/O pins that can be routed to the mezzanine I/O pins on the profile above. There are proprietary solutions, such as Acromag's AcroPack®[10], and small form factor mezzanines such as Mini-PCI Express[11] or M.2[12] could be used with custom cabling solutions. However, the recently formed VITA 93 Small Form Factor Mezzanine (SFFm) working group is working to define such a form factor. It is expected that the SFFm will provide the same industry-standard mezzanine I/O expansion capability for VNX+ that OpenVPX enjoys with VITA 42[13] and 61[14] XMC modules.

COMMUNICATIONS PROTOCOLS

SOSA also defines communications protocols that are approved for the communications ports. As-of SOSA Technical Standard v2 Snapshot 1, SOSA has not defined protocols for VNX+ plug-in cards. However, SOSA does define protocols for VPX plug-in cards, and VITA 90.0 captures those same protocols, including:

- Data Plane
    - 10GBASE-KR

    - 40GBASE-KR4
    - 25GBASE-KR
    - 100GBASE-KR4
- Control Plane
    - 1000BASE-KX
    - 10GBASE-KR
    - 25GBASE-KR
- Expansion Plane
    - PCIe Gen2, Gen3, and Gen4
    - Aurora
    - Serial Front Panel Data Port (Serial FPDP)
    - LVDS discretes

Therefore, it can be expected that upcoming releases of the SOSA Technical Standard will flesh out the approved protocols for the VNX+ communications planes, along with any new ports defined in any as-yet unannounced slot profiles. In addition, space-specific protocols are expected to be adopted once the space-specific profiles are defined and added to the standard.

POWER, COOLING, AND ALTERNATE MECHANICAL STRUCTURES

While not yet discussed in depth in the SOSA Technical Standard, the draft VITA 90 documents either include or have on their roadmap for inclusion a number of other important features, including power supply plug-in cards, thermal performance data, 39mm module width, and alternate mechanical structures featuring wedge locks and direct-to-surface mounting.

As these topics mature in the VITA 90 drafting efforts, it can be expected that SOSA will address them in a future version of the Technical Standard.

VNX+ AND AUTONOMOUS VEHICLES

There are a number of factors that make VNX+ attractive to autonomous vehicles, either as a vehicle control platform or as sensor payload. Numerous elements of VNX+ and technical details of the SOSA Technical Standard have direct applicability to autonomous systems, including:

- Compact physical size
- Naturally rugged design
- Modularity and scalability
- High-speed board-to-board communications channels
- Built-in I2C and SPI support (for peripheral device control)
- Support for coax and fiber optic links
- Scalable processing capacity up to 25W/slot (or more, depending on the cooling infrastructure)

As autonomous vehicles are usually smaller than crewed vehicles, there is a natural attraction to VNX+'s compact size and its naturally rugged conduction-cooled design. Taking only about 30% of the per-slot volume of a 3U OpenVPX card, VNX+ can be packaged and deployed in applications simply too small for 3U VPX. Good examples of this include Common Launch Tube deployed UAVs like Raytheon's Coyote[15] or Area-I's ALTUS-600M[16]. Small UUVs such as HII's Remus[17], robotic ground vehicles like General Dynamics Land Systems Multi-Utility Tactical Transport (MUTT)[18], and naval autonomous surface vehicles or small combat craft are also good candidates for VNX+ deployment, although the latter two often do not have the severe SWAP constraints that drive UAV or UUV applications to small form factors such as VNX+. There are examples that are good candidates, however, such as SubSeaSail's Gen7[19] (Reference 19).

VNX+ is designed for rugged applications where wide temperature ranges, shock and vibration, and other environmental factors are the norm. Thus, no additional hardening of the electronics is needed for these deployed environments.

The Unique External I/O (UEIO) region of the S0 Utility Segment also provides a built-in capability for direct vehicle control through I2C or SPI-based controllers without adding mezzanine I/O. As a result, VNX+ has a capability for external device control that other form factors like OpenVPX lacks. Should other interfaces like MIL-STD-1553, ARINC 429, or CANBus be required (although the latter can easily be implemented with a SPI/CANBus interface chip), mezzanine I/O boards such as the new VITA 93 Small Form Factor Mezzanine could be added, who's I/O signals will be mapped to backplane pins in several VNX+ Payload Profiles.

The biggest strength for VNX+, however, is its inclusion in the SOSA Technical Standard. That will give VNX+ an established industry-approved reference architecture that can be leveraged to quickly integrate complex sensor payloads and control electronics using off-the-shelf building blocks like modules and backplanes that simply "just work". The standard not only enables rapid integration of new systems, but it enables technology insertion for either technology upgrades or for quickly creating new derivative plat leveraging new, innovative products.

SUMMARY

In summary, there are a number of elements of VNX+ that should prove attractive to autonomous vehicles, regardless of their employment (surface, air, sea surface, or undersea):

- Size: The compact size of VNX+ makes it immediately attractive to autonomous vehicles, which often have severe size constraints.

- Scalability: VNX+ is, by its nature, a scalable form factor, with switch cards defined that form its multi-plane switched networks.

- Standards-based: The fact that VNX+ is based on an international standard (VITA 90), and that the SOSA Technical Standard (itself freely available for use internationally) has adopted it, means that a market for products based on the standard is very likely to develop, with multiple suppliers and products to choose from. The SOSA standard strongly drives interoperability, so integrators can be assured that products from different suppliers will "just work".

- Ruggedness: VNX+ is designed for rugged environments and is, by its nature, a conduction-cooled form factor. This means that specially-designed cooling frames are not needed – only a cold wall on two or three sides of the modules is required. This simplifies system designs.

- UEIO: VNX+'s UEIO port introduces a standardized device control capability missing from many form factors. This enables control functionality without the need for additional I/O boards.

- High-Performance: VNX+ will have natural performance limits due to the thermal constraints of its small size, but it provides extremely high potential bandwidth between slots, making it ideal for complex sensor processing applications.

- Coax and Fiber Optic Links: VNX+ aperture fill blocks provide the means to directly connect to sensors or high-speed communications channels without taking up additional slots.

- Out-of-band System Management: To provide for system monitoring and control, and providing a mechanism for system resiliency.

Only time will tell whether VNX+ proves to be attractive to the autonomous vehicle community, but the potential certainly exists due to the hard work of the VITA 90 Working Groups and the SOSA Small Form Factors Subcommittee who have devoted such time and effort into crafting such a fully defined standard.

REFERENCES

[1] VITA "ANSI/VITA 74.0-2022 Compliant System Small Form Factor Module (VNX) Base Standard", November 3, 2022 revision of "ANSI/VITA 74.0-2017", December 4, 2017; refer to: www.vita.com/Standards

[2] VITA "ANSI/VITA 46.0-2019 VPX Baseline Standard", February 1, 2019; refer to: www.vita.com/Standards

[3] VITA "ANSI/VITA 65.0-2021 OpenVPX System Standard", October 4, 2021; refer to: www.vita.com/Standards

[4] https://www.opengroup.org/sosa

[5] The Open Group "Technical Standard for SOSA™ Reference Architecture, Edition 2.0, Snapshot 1", June 2022, https://publications.opengroup.org/s221

[6] VITA "ANSI/VITA 66.0-2016: Optical Interconnect on VPX – Base Standard", July 1, 2016; refer to: www.vita.com/Standards

[7] VITA "ANSI/VITA 67.0-2019: Coaxial Interconnect on VPX – Base Standard", April 9, 2019; refer to: www.vita.com/Standards

[8] VITA "ANSI/VITA 67.3-2020: Coaxial Interconnect on VPX, Spring-Loaded Contact on Backplane", January 21, 2021; refer to: www.vita.com/Standards

[9] Samtec "SEAF-RA/SEAM - High Speed Characterization Report", October 19, 2021. https://sudden-docs.samtec.com/testreports/20211019_hsc-report-sma_seam-02_seaf-ra_10row.pdf?_gl=1*1mr9bdf*_ga*NDQ0NDgxODg1LjE2NzYwNTQ2NDk.*_ga_3KFNZC07WW*MTY4MDg4Nzk4NC4xNi4xLjE2ODA4ODg3MzguMzcuMC4w

[10] https://www.acromag.com/product-category/embedded-i-o-processing-solutions/acropack-mini-pci-express-based-interface-i-o-boards/acropack-carriers/

[11] PCI-SIG "PCI Express® Mini Card Electromechanical Specification Revision 2.1", December 15 2016; refer to https://pcisig.com/specifications

[12] PCI-SIG "PCI Express M.2 Specification Revision 4.0, Version 1.1", November 9, 2022; refer to https://pcisig.com/specifications

[13] VITA "ANSI/VITA 42.0-2021 XMC: Switched Mezzanine Card (XMC) Auxiliary Standard", September 30, 2021; refer to: www.vita.com/Standards

[14] VITA "ANSI/VITA 61.0-2022 XMC 2.0", February 10, 2022; refer to: www.vita.com/Standards

[15] https://www.raytheonmissilesanddefense.com/what-we-do/counter-uas/effectors/coyote

[16] https://areai.com/altius-600-2/

[17] https://hii.com/capabilities/remus-uuvs/

[18] https://www.gdls.com/s-met/

[19] https://subseasail.com/unmanned-autonomous-surface-vessels/our-tech

MECHANISM-FREE "SOLID-STATE" ROTORS FOR DRONES, UAVS, WIND AND HYDROKINETIC TURBINES

Onur Bilgen* **and Bharg Shah†**

This paper presents a summary of the ongoing research on a mechanism-free piezo-composite rotor concept. The proposed hub-rotor system can be implemented in drones and unmanned-aerial-vehicles such as single-rotor, tandem-rotor, multi-copter, and ducted-fan rotorcraft, or other rotating systems such as wind and hydrokinetic turbines, turbine engines, and marine propellers. The preliminary feasibility of the so-called solid-state rotor concept has been previously demonstrated through flight-testing a prototype on a quad-copter vehicle platform. This paper presents a summary of previous research on three different prototypes.

INTRODUCTION

In this research, the authors extend their concepts on the mechanism-free solid-state fixed-wing aircraft and piezo-composite ornithopters [1,2,3,4,5], and present ongoing research on the so-called *Direct-Drive Solid-State Rotor*. The ongoing research objective is to design, optimize, and demonstrate a fully functional direct-drive solid-state piezo-composite rotor for use in various applications. To this end, several electric-driven hub-rotor systems have been proposed, developed and successfully tested [6,7,8,9,10,11,12,13,14,15]. Specifically, it has been shown that the proposed concept can be implemented in small drones or unmanned-aerial-vehicles (UAVs) such as single-rotor, tandem-rotor, multi-copter, and ducted-fan rotorcraft.

THE MECHANISM-FREE "SOLID-STATE" ROTOR CONCEPT

A mechanism-free solid-state hub-rotor system with shape changing blades and electromagnetically powered rotating electronics has been proposed by the authors. A solid-state or semi-solid-state strain-inducing actuator is proposed to change/control the shape of the rotor blade, wind-turbine or engine-turbine blade, or for any type of rotating device used to interact with surrounding fluid. Strain-induced actuation can be accomplished easily with piezo-composite actuators, shape-memory alloy materials, or with any smart material with electrical to mechanical transduction property. The electronics are proposed to be on the rotating system, where the power and data transmission is achieved without contact using electromagnetic transmission. An illustration of the proposed concept is presented in Figure 1.

* Assistant Professor, Mechanical and Aerospace Engineering, Rutgers University, Piscataway, NJ 08854.
† Graduated M.S. Student, Mechanical and Aerospace Engineering, Rutgers University, Piscataway, NJ 08854.

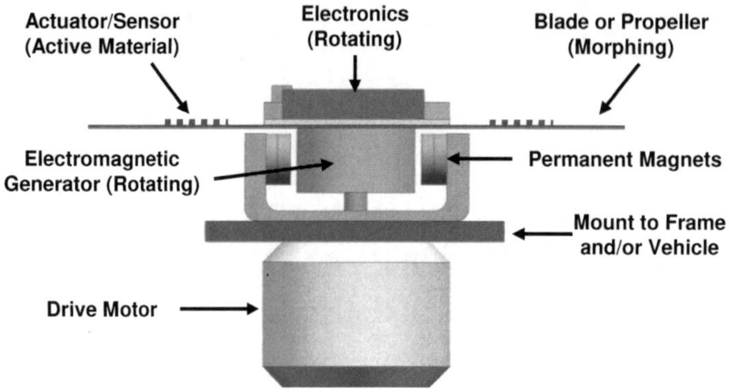

Figure 1: An illustration of the proposed concept where the electronics and the propeller are on the same side of the drive shaft.

An alternative assembly is shown in Figure 2 where the electronics and the propeller are on the opposite ends of the drive shaft and the drive motor.

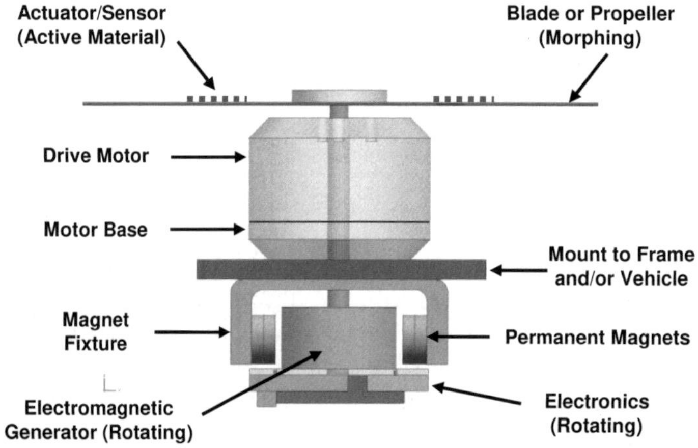

Figure 2: An illustration of the proposed concept in an alternative assembly where the electronics and the propeller are on the opposite ends of the drive shaft and the drive motor.

The configurations shown above lead to a fully contained mechanism-free rotor-hub system with built-in electronics with energy harvesting and wireless data transmission capabilities. This rotor-hub system can be fully sealed enabling operation in corrosive, conductive or otherwise undesirable fluids and mixtures. In Figure 1 and Figure 2, the rotor-hub system is mechanically driven by an "outrunner" type motor by the use of a drive shaft. A generator coil driven by the drive shaft is used to generate alternating current as it rotates in the magnetic field generated by the fixed permanent magnets. The generator output is connected to a circuit board, which hosts circuitry to condition the power and control the surface-bonded or embedded active material devices. A wireless receiver interprets control signals sent by a radio transmitter that is located on the vehicle and connected to the flight control unit (i.e., autopilot).

A type of piezocomposite device known as the Macro-Fiber Composite (MFC) actuator has been increasingly used since its invention in 2000 to substitute for conventional mechanical actuators in aerodynamic control surfaces. Conformal MFC actuators have been successfully implemented in small aircraft by various researchers [16,17,18,19,20,21,22,23]. Publications in this area have demonstrated the feasibility of using piezoceramic actuators in small fixed-wing aircraft [16, 17,20,21,22]. Consequently, the MFC actuators are utilized as the active element for the prototypes presented in this research.

DEMONSTRATION OF FEASIBILITY THROUGH FLIGHT TESTS

An operational quadcopter has been developed where all four rotors are implemented with the proposed solid-state hub-rotor system [10,11,14]. It is noted that the proposed solid-state hub-rotor system is not specific to a quadcopter or a rotary-wing aircraft; however, a quad-copter provides a platform where the implementation is relatively easy compared to other platforms. The experimental quad-copter is shown in Figure 3.

Figure 3: The experimental quad-copter with four rotors equipped with the hub-rotor system with piezocomposite blades and rotating electronics powered with an electromagnetic generator. The vehicle a) on ground, and b) in flight.

Several benchtop and flight tests showed operation of the proposed concept during flight. Both the main control of the quad-copter and the manual control of the piezoelectric materials on the rotors are achieved by a single 8-channel transmitter. All four rotors and the corresponding receivers on the hubs are connected to the sixth auxiliary channel allowing the sixth channel on the transmitter to control all piezocomposite blades collectively. A jumper-selector on each hub allows the rotors to be in collective, asymmetric, or individual modes. This arrangement is chosen only for experimental purposes – the number and arrangement of control channels between the transmitter and the receiver is highly dependent on the application and the function necessary.

FIRST PROTOTYPE – A MODIFIED COMMERCIAL PROPELLER

A compact flight-ready (and later flight-tested) hub-rotor prototype is developed [14]. Figure 4 shows one of the prototypes developed mounted on an outrunner-type brushless motor. The electronics and the generator are mounted on the opposite side of the motor, below the fixed frame. The entire hub-rotor system is mounted on a fixture to enable thrust measurement, which will be discussed later. The wireless transmitter is not shown in the figure.

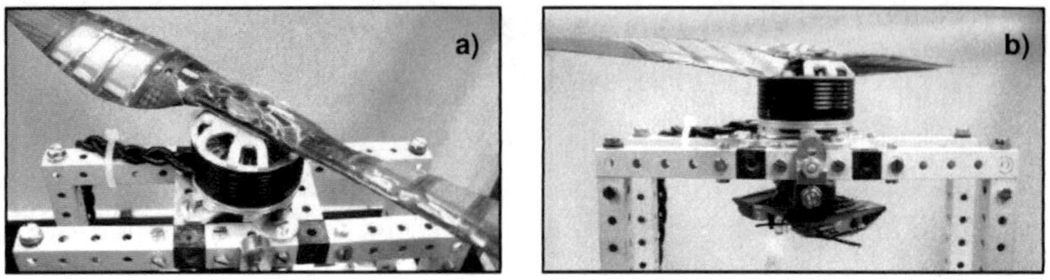

Figure 4: A prototype of the a) piezocomposite blades mounted on an outrunner-type brushless motor, and b) the electronics and the generator mounted on the opposite side of the motor. The rotor diameter is 406 mm (16 inch.)

Three MFC actuators are used per blade. The composite blades are modified to reduce their stiffness in the chordwise direction while minimizing the effect of the modification on their bending and twist rigidity. Figure 5 shows the operation of the complete system at a speed of approximately 4715 rpm. When the control signal is changed from 0% to 100%, a red LED indicator on the electronics board, in parallel with the DC-DC converter, turns on confirming the operation of the system.

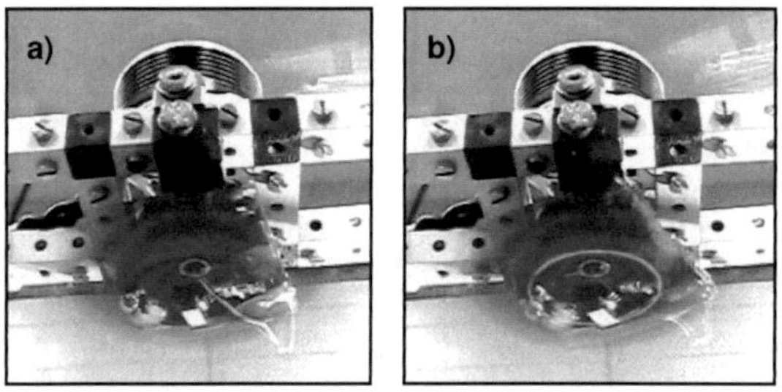

Figure 5: Observation of the operation of the hub-rotor system using a strobe light. The control signal is a) turned off at approximately 4715 rpm, and b) turned on at approximately 4680 rpm.

In addition to the LED indication, the motor speed reduces from 4715 rpm to approximately 4680 rpm when the DC-DC generators are powered. This is as expected since powering the electronics induces electromagnetic drag on the drive motor.

Benchtop Static-Thrust Experiments – Prototype I

A set of experiments are conducted to determine the change in thrust force when the piezocomposite blades are actuated [14]. The prototype discussed in the previous section is placed on an electronic scale with a strain-gauge based load cell, and a National Instruments (NI) data acquisition (DAQ) system is used to automate the experiments. The experimental setup is illustrated in Figure 6.

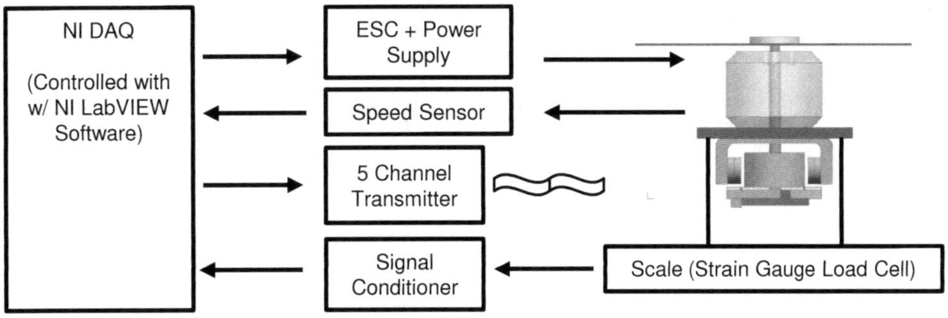

Figure 6: A functional illustration of the bench-top test apparatus including the hub-rotor system mounted on a load scale through a rigid frame.

An NI LabVIEW based program is used to control the pulse-position-modulation (PPM) signal sent to a 5-channel transmitter, which is then interpreted by the received on the rotating electronics. A PWM signal is sent to an electronic-speed-controller (ESC). The power to the ESC is supplied by a precision DC power supply set to an output of 20 Vdc. The ESC then routes the power to the brushless out-runner type motor. The thrust force is measured by the strain-gauge load cell, and the voltage output is balanced and gained using a strain-gauge signal conditioning amplifier. The bench-top test apparatus including the prototype of the hub-rotor system mounted on the load scale through a rigid frame is shown in Figure 7. The electronics used in the experiment are visible in the background.

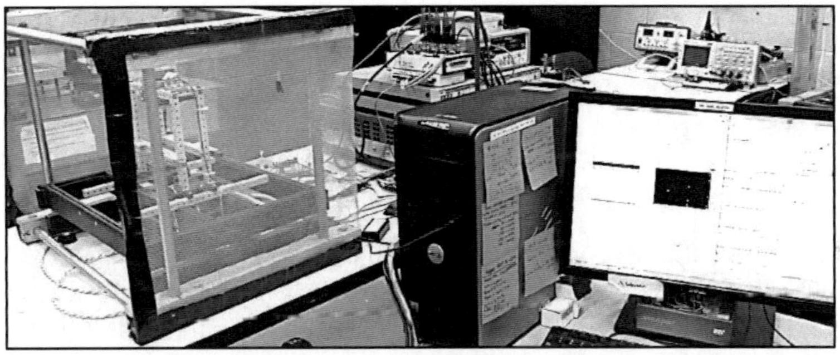

Figure 7: Bench-top test apparatus including a prototype of the hub-rotor system mounted on a load scale through a rigid frame. The prototype and the electronic scale are surrounded with a wire screen for safety purposes.

The sweep of the rotor sweep is conducted in an increasing-decreasing order and averaged to minimize the effects of hysteresis inherent in the system. For each test case, 20 seconds of data is collected after steady state is reached. The time-averaged thrust force is calculated from the experimental measurements by:

$$T_{AVE} = \frac{1}{N}\sum_{i=1}^{N} T_i \tag{1}$$

The propeller thrust coefficient is calculated with the well-known formula

$$C_T = \frac{T}{\rho f^2 D^4} \tag{2}$$

where ρ is the density of air, f is the measured angular frequency in rev/sec, and D is the rotor disk diameter. Since the angular frequency varies during experiments, it is measured and used to normalize lift force to result in the thrust coefficient. Assuming that density of air and the area of the propeller disk remains constant, the thrust coefficient quantifies the change in aerodynamic efficiency of the propeller purely due to the change of its shape. Figure 12 presents the experimentally measured thrust coefficient as a function of rotor angular speed measured in RPM.

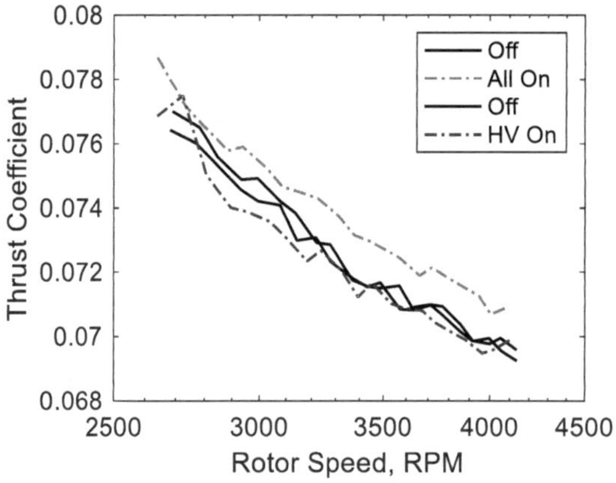

Figure 8: Prototype I benchtop experimental results for camber control turned off and on: a) measured thrust, and b) the normalized thrust metric.

In the figure, four different test conditions are presented. The "Off" condition is where the high-voltage (HV) amplifier is turned off. The "HV On" condition indicates the test where the high-voltage amplifier is turned on; however, the output is not connected to the MFCs on the propeller. In this state, the MFCs are in a short-circuit condition. The "All On" condition indicates where the HV amplifier is on and connected to the MFCs on the propeller.

SECOND PROTOTYPE – A CUSTOM PROPELLER WITH SYMMETRIC PROFILE

Figure 4 shows the so-called second prototype mounted on an outrunner-type brushless motor [10]. The electronics and the generator are mounted on the opposite side of the motor, below the fixed frame. The entire hub-rotor system is mounted on a fixture to enable thrust measurement. Three MFC actuators are used per blade.

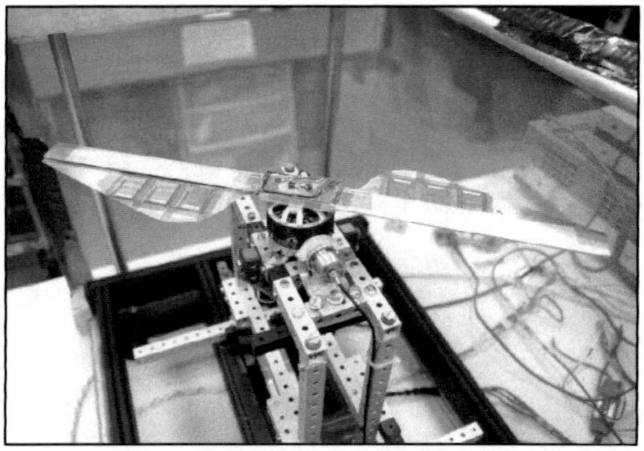

Figure 9: Prototype II with piezocomposite blades mounted on an outrunner-type brushless motor along with electronics and the generator mounted on the opposite side of the motor. The rotor diameter is 406 mm (16 inch.)

Benchtop Static-Thrust Experiments – Prototype II

A set of experiments are conducted to determine the change in thrust force when the piezocomposite blades are actuated [10]. The prototype is placed on the electronic scale (discussed earlier) with a strain-gauge based load cell, and the National Instruments (NI) data acquisition (DAQ) system is used to automate the experiments. The sweep of the rotor speed is conducted in an increasing-decreasing order and averaged to minimize the effects of hysteresis inherent in the measurement system. For each test case, 20 seconds of data is collected after steady state is reached. The time-averaged thrust force is calculated from experimental measurements. Since the angular frequency varies during experiments, it is measured and used to normalize lift force to result in the thrust coefficient.

Figure 12 presents the experimentally measured thrust coefficient as a function of rotor angular speed measured in RPM. Assuming that density of air and the area of the propeller disk remains constant, the thrust coefficient quantifies the change in aerodynamic efficiency of the propeller purely due to the change of its shape.

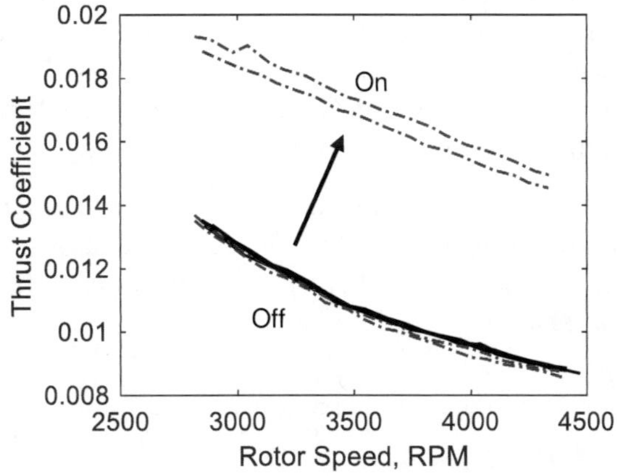

Figure 10: Prototype II thrust coefficient from benchtop experiments for the control signal off and on.

In the figure, three different test conditions are presented. The "Off" condition indicates either the high-voltage (HV) amplifier is turned off, or where HV amplifier is turned on; however, the output is not connected to the MFCs on the propeller. In this state, the MFCs are in a short-circuit condition. The "On" condition indicates where the HV amplifier is on and connected to the MFCs on the propeller.

THIRD PROTOTYPE – A CUSTOM PROPELLER WITH CAMBERED PROFILE

Figure 4 shows the so-called third rotor prototype mounted on an outrunner-type brushless motor on a static thrust test stand [15]. The electronics and the generator are mounted on the opposite side of the motor, below the fixed frame. The entire hub-rotor system is mounted on a fixture to enable thrust measurement. Three MFC actuators are used per blade.

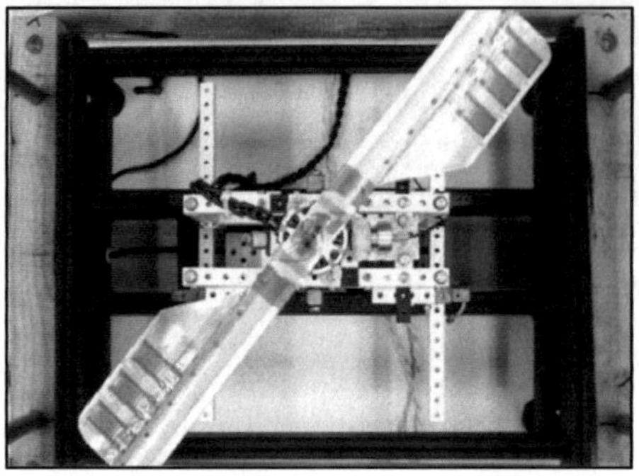

Figure 11: Prototype III with piezocomposite blades mounted on an outrunner-type brushless motor along with electronics and the generator mounted on the opposite side of the motor. The rotor diameter is 406 mm (16 inch.)

Benchtop Static-Thrust Experiments – Prototype III

A set of experiments are conducted to determine the change in thrust force when the piezocomposite blades are actuated [15]. The prototype is tested using the same system described above for the other prototypes. The sweep of the rotor sweep is conducted in an increasing-decreasing order and averaged to minimize the effects of hysteresis inherent in the system. For each test case, 20 seconds of data is collected after steady state is reached.

Figure 12 presents the experimentally measured thrust coefficient as a function of rotor angular speed measured in RPM. Assuming that density of air and the area of the propeller disk remains constant, the thrust coefficient quantifies the change in aerodynamic "efficiency" of the propeller purely due to the change of its shape.

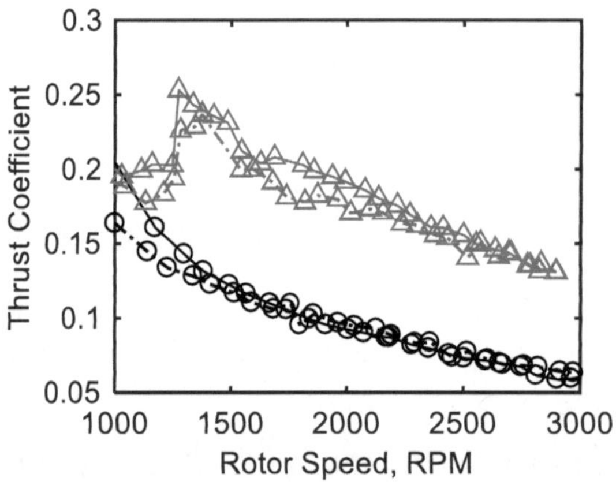

Figure 12: Prototype III thrust coefficient from benchtop experiments for the control signal off (black markers) and on (red markers).

In the figure, two different test conditions are presented. The black line represents the response when the high-voltage amplifier is turned off. The red line represents the response when the high-voltage amplifier is turned on.

CONCLUSIONS

In this paper, the ongoing development of a novel mechanism-free solid-state hub-rotor system is summarized. Flight tests have been previously conducted to show the operation of the rotor during flight and in realistic environmental conditions. Consequently, the feasibility of a solid-state rotor using smart materials has been demonstrated. This paper presented a summary of benchtop experimental results from three different prototypes.

ACKNOWLEDGMENTS

The authors acknowledge the support of the following: Old Dominion University Equipment Trust Fund, Frank Batten College of Engineering and Technology, Rutgers University School of Engineering, and Rutgers University Department of Mechanical Engineering.

REFERENCES

[1] Bilgen, O., Friswell, M. I., and Taqiuddin, M., "Coupled Modeling and Optimization of Piezocomposite Wings." ASME 2013 Conference on Smart Materials, Adaptive Structures and Intelligent Systems (SMASIS2013), September 16-18, 2013, Snowbird, Utah, USA, SMASIS2013-3311, 2013.

[2] Hauris, F. and Bilgen, O., "Induced Strain Actuation for Solid-State Ornithopters: A Parametric Study," ASME 2014 Conference on Smart Materials, Adaptive Structures and Intelligent Systems (SMASIS2014), September 8-10, 2014, Newport, Rhode Island, USA, SMASIS2014-7527, 2014.

[3] Bounthisane, R., Gumapas, E. and Bilgen, O., "Solid-State Ornithopter: A Feasibility Study," 23rd AIAA/AHS Adaptive Structures Conference (SciTech 2015) American Institute of Aeronautics and Astronautics, January 5-8, 2015, Kissimmee, Florida, USA, 2015. doi: 10.2514/6.2015-1723

[4] Hauris, F., and Bilgen, O., 2016, "Induced Strain Actuation for Solid-State Ornithopters: A Geometric Configuration Study," 57th AIAA/ASCE/AHS/ASC Structures, Structural Dynamics, and Materials Conference (AIAA SciTech), AIAA 2016-0818, 4-8 January 2016, San Diego, California, USA.

[5] Hauris, F. and Bilgen, O., 2016, "Induced Strain Actuation for a Solid-State Wing," ASME 2016 Conference on Smart Materials, Adaptive Structures and Intelligent Systems (SMASIS2016), September 28-30, 2014, Stowe, Vermont, USA, SMASIS2016-9020.

[6] Shah, B., and Bilgen, O., 2022, "Aerodynamic Modeling and Analysis of a Variable Camber Piezocomposite Rotor," ASME 2022 Conference on Smart Materials, Adaptive Structures and Intelligent Systems (SMASIS2022), Dearborn, Michigan, September 12 - 14, 2022, SMASIS2020-91111.

[7] Zhang, K., and Bilgen, O., 2020, "Multi-Fidelity Aerodynamic Modeling of A Floating Offshore Wind Turbine Rotor," ASME International Mechanical Engineering Congress and Exposition (IMECE2020), November 15-19, 2020, Virtual.

[8] Shah, B., and Bilgen, O., 2020, "A Piezocomposite Actuated Propeller with an Onboard Power Generator," ASME 2020 Conference on Smart Materials, Adaptive Structures and Intelligent Systems (SMASIS2020), September 14-16, 2020, Irvine, CA, USA, SMASIS2020-15853.

[9] Ferreira de Castro, T.C., and Bilgen, O., 2019, "A Piezocomposite Solid-State Rotor: Theoretical Analysis Of Thrust And Efficiency Metrics," ASME 2019 Conference on Smart Materials, Adaptive Structures and Intelligent Systems (SMASIS2019), September 9-11, 2019, Louisville, KY, USA, SMASIS2019-5558.

[10] Bilgen, O., 2017, "A Piezocomposite Rotor with Contactless Power and Data Transmission," ASME 2017 Conference on Smart Materials, Adaptive Structures and Intelligent Systems (SMASIS2017), September 18-20, 2017, Snowbird, Utah, USA, SMASIS2017-3733.

[11] Bilgen, O., and Alberts, T. E., "Solid-State Rotor: A Demonstration," 24th AIAA/AHS Adaptive Structures Conference (SciTech 2016) American Institute of Aeronautics and Astronautics, January 4-7, 2016, San Diego, California, USA, 2016. doi: 10.2514/6.2016-1080

[12] Bilgen, O., and Alberts, T. E., 2016, "A Solid-State Rotor for Drones," Poster Presentation with Paper, AUVSI Exponential 2016, New Orleans, LA, USA.

[13] Bilgen, O., and Alberts, T. E., 2016, "A Direct-Drive Piezocomposite Solid-State Rotor with Contactless Power and Data Transmission," ASME 2016 Conference on Smart Materials, Adaptive Structures and Intelligent Systems (SMASIS2016), September 28-30, 2014, Stowe, Vermont, USA, SMASIS2016-9021.

[14] Bilgen, O., 2017, "Solid-State Propulsion for Rotary- and Fixed-Wing Aircraft," Poster Presentation with Paper, AUVSI Exponential 2017, Dallas, TX, USA.

[15] O. Bilgen, "Solid-state propulsion for rotary- and fixed-wing aircraft," presented at the AUVSI XPONENTIAL 2018, Colorado Convention Center, Denver, CO, 1 May 2018 - 3 May 2018, 2018.

[16] Bilgen, O., Kochersberger, K. B., Diggs, E. C., Kurdila, A. J. and Inman, D. J., "Morphing Wing Micro-Air-Vehicles via Macro-Fiber-Composite Actuators," 48th AIAA/ASME/ASCE/AHS/ASC Structures, Structural Dynamics, and Materials Conference, Honolulu, HI, AIAA-2007-1785, 2007.

[17] Bilgen, O., Kochersberger, K. B. and Inman, D. J., "Macro-Fiber Composite Actuators for a Swept Wing Unmanned Aircraft," *Aeronautical Journal*, Special issue on Flight Structures Fundamental Research in the USA, publication of Royal Aeronautical Society, Vol. 113, No. 1144, 2009.

[18] Paradies, R. and Ciresa, P., "Active wing design with integrated flight control using piezoelectric macro fiber composites," *Smart Materials and Structures*, Vol. 18, 2009. doi:10.1088/0964-1726/18/3/035010

[19] Wickramasinghe, V. K., Chen, Y., Martinez, M., Kernaghan, R. and Wong, F., "Design and Verification of a Smart Wing for an Extremely-Agile Micro-Air-Vehicle,", 50th AIAA/ASME/ASCE/AHS/ASC Structures, Structural Dynamics, and Materials Conference, Palm Springs, CA, AIAA 2009-2132, 2009.

[20] Butt, L., Bilgen, O., Day, S., Sossi, C., Weaver, J., Wolek, A., Inman, D.J. and Mason, W.H., "Wing Morphing Design Utilizing Macro Fiber Composite," Society of Allied Weight Engineers (SAWE) 69th Annual Conference, Virginia Beach, VA, Paper Number 3515-S, 2010.

[21] Bilgen, O., Butt, L.M., Day, S.R., Sossi, C.A., Weaver, J.P., Wolek, A., Mason, W.H. and Inman, D.J., "A Novel Unmanned Aircraft with Solid-State Control Surfaces: Analysis and Flight Demonstration," 52nd AIAA/ASME/ASCE/AHS/ASC Structures, Structural Dynamics, and Materials, Denver, CO, 2011.

[22] Bilgen, O., Butt, L.M., Day, S.R., Sossi, C.A., Weaver, J.P., Wolek, A., Mason, W.H. and Inman, D.J., "A Novel Unmanned Aircraft with Solid-State Control Surfaces: Analysis and Flight Demonstration," *Journal of Intelligent Material Systems and Structures*, Vol. 24, No. 2, 2012, pp. 147–167. doi: 10.1177/1045389X12459592

[23] Ohanian, O. J., Hickling, C., Stiltner, B., Karni, E. D., Kochersberger, K. B., Probst, T., Gelhausen, P. A. and Blain, A. P., "Piezoelectric Morphing versus Servo-Actuated MAV Control Surfaces," 53rd AIAA/ASME/ASCE/AHS/ASC Structures, Structural Dynamics and Materials Conference, Honolulu, HI, AIAA 2012-1512, 2012.

AUTONOMOUS EXPLORATION AND MAPPING OF UNKNOWN ENVIRONMENTS WITH UNMANNED AERIAL SYSTEMS (UAS)

Eric Thorn, Ph.D.*

Southwest Research Institute (SwRI) has developed technologies to enable unmanned aerial systems (UAS) to autonomously explore and map unknown environments. The presentation will summarize the development of simultaneous localization and mapping (SLAM), world modeling, and exploration-based planning algorithms to facilitate navigation in and mapping of these environments. The presentation will also include a summary of experiments conducted using the novel technologies to map several environments. First, we will discuss results from the European Robotics Hackathon (EnRicH) held in late 2021. This event was held at the AKW Zwentendorf nuclear power plant in Zwentendorf Austria. During this event SwRI's UAS autonomously navigated and mapped the interior of several parts of the power plant and detected live radiation sources. We will then summarize efforts to autonomously explore and map several underground caves near our headquarters in San Antonio, TX.

INTRODUCTION

This paper summarizes the development of a custom unmanned aerial system (UAS) platform and software architecture that enables autonomous exploration and mapping of unknown environments. By its nature, exploration of unstructured and unknown environments does not afford a system the same advanced knowledge of the operating environment (i.e., high-fidelity *a priori* maps). A system must instead rely solely on its onboard sensors to map, localize, and navigate. This is especially important in indoor environments where global positioning system (GPS) updates may be unreliable or absent altogether. These capabilities offer significant benefits for public safety and defense applications, among others, such as exploring and mapping industrial facilities, underground caves and tunnels, and damaged or contested buildings.

The paper is organized as follows:

- Platform – summarizes platform and sensing and other hardware
- Localization and Mapping – summarizes methods used to localize and perform mapping
- World Modeling – summarizes methods used to model the operating environment and enable exploration
- Navigation – summarizes algorithms used to perform global and local planning as well as path following
- Testing and Experimentation – summarizes testing and experimentation in real-world environments

* Manager R&D – Robotic Autonomy Section, Intelligent Systems Division, Southwest Research Institute.

PLATFORM

A commercially available Tarot 650 UAS, shown in Figure 1, was used for the development and experimentation covered in this paper. A Pixhawk 4 with integrated barometer, magnetometer, and inertial measurement unit (IMU) and open-source PX4 flight control solution. For perception, an Ouster OS0-64 lidar was integrated to provide medium-to-long range point cloud data to support mapping and obstacle avoidance, along with a Tarabee ranger time-of-flight (ToF) sensor to detect objects directly below the UAS. A set of ultrasonic sensors was also integrated to help fill in near-field gaps and to specifically detect clear surfaces such as windows. Additionally, a commercial radiation sensor was integrated to support detection and mapping of radioactive sources. For computation, an Intel NUC with integrated Intel i7-1165G7 processor and Odroid C4 were used. The NUC handled processing of lidar point cloud data, while the Odroid managed navigation and planning processes, along with vehicle health, communication, and path following control.

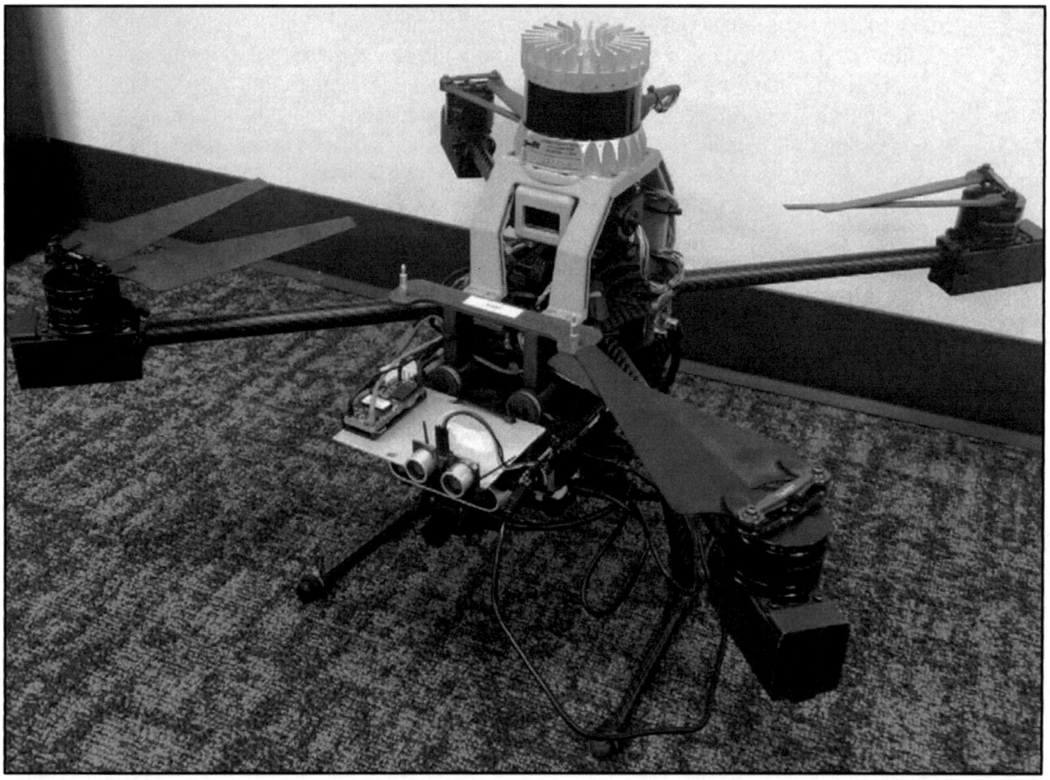

Figure 1. Tarot 650 Development Platform.

LOCALIZATION AND MAPPING

The Google Cartographer open-source simultaneous localization and mapping (SLAM) library was integrated and evaluated to provide SLAM in three dimensions. Three range sensors (RealSense D435i camera system, Velodyne VLP-16 lidar, and Ouster OS0-64 lidar) were evaluated for suitability to provide the necessary point cloud data to Cartographer. Cartographer* is a set of open-source libraries written for use with both 2D and 3D SLAM. While Cartographer is complex, the pieces of the system may be distilled down to four main components: input data, the pose extrapolator, local SLAM, and global SLAM. Cartographer may receive data from a variety of inputs,

* https://google-cartographer.readthedocs.io/en/latest/

including point clouds, laser scans, GPS, odometry, IMUs, and landmarks. High frequency input is preferred since Cartographer performs motion estimation; therefore, receiving data quickly increases the quality of the SLAM solution.

The input data is passed into the local SLAM portion of the algorithm. This portion is responsible for building up the submaps of a map. A submap is accumulated over distance and data. As Cartographer receives range data, it extrapolates it based on the current motion of the vehicle to make an initial guess as to where the data should be in the submap. A scan matcher is used to match the range data to the previous scans in the submap. Figure 1 shows sample scan matching data for multiple lidar configuration: a) full revolutions at 20 Hz and b) individual packets at 1,280 Hz. Once matched, a motion filter examines whether the distance, angle, or time between the current scan and the latest scan in the submap is significant enough before adding it to the submap. This process is repeated until the submap has reached a preconfigured size. Once a submap is completed, it is set aside to be used for loop closure constraints as the UAS continues its trajectory. Note that while a submap should have minimal drift, it is expected that it will drift over time without loop closure.

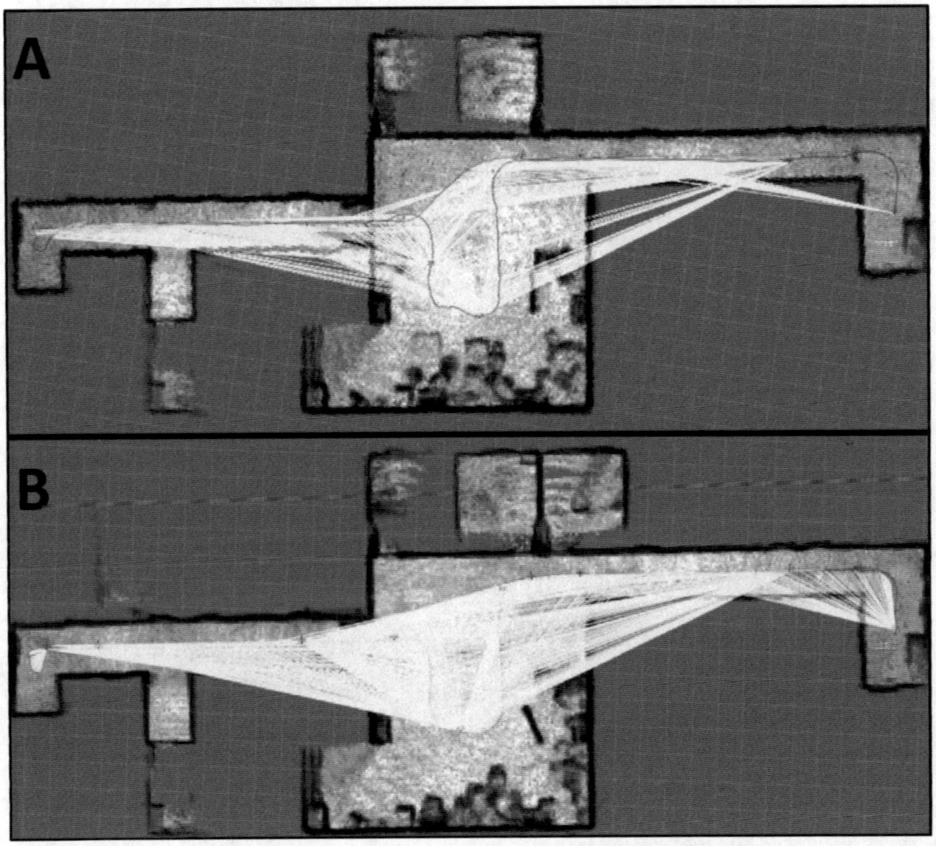

Figure 2. Sample Cartographer Scan Matches.

The global SLAM portion of the algorithm calculates loop closure constraints and adjusts to create the global map. Loop closure is where Cartographer recognizes an area it has seen before and removes any drift that has accumulated since it was last in that position. Cartographer searches for loop closures between a newly finished submap and previous submaps that are within a search window. The global SLAM algorithm uses the loop closure constraints to optimally rearrange the submaps into a global map. Figure 2 and Figure 3 show sample 2D and 3D maps, respectively, generated for a test flight conducted in a large, indoor, warehouse-like test environment.

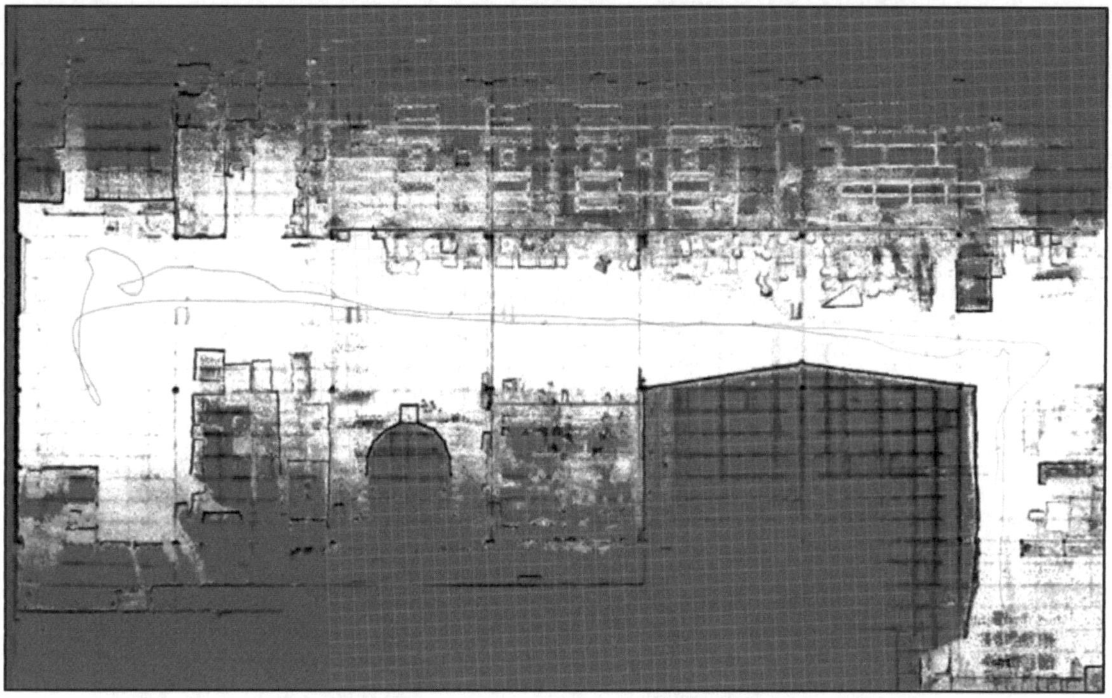

Figure 3. Sample Cartographer 2D Map.

Figure 4. Sample Cartographer 3D Map.

WORLD MODELING

To support long-term route planning applications such as exploration in an unknown area, a three-dimensional world modeling framework was implemented. The world model was developed to ingest various perception sensor data to support mapping large spaces while not overloading computational resources. Two world model layers were developed and utilized: a frontier layer and an obstacle layer.

The frontier layer looks at the available map generated by Cartographer and detects frontiers, or unexplored areas, assigning a weight to each area. The obstacle layer represents potential obstacles around the UAS. The frontier layer charts an array of available frontiers with assigned weights. It is primarily used by exploration planners to determine new areas to explore. Frontier exploration is a technique used to gain new information about the world by moving from the boundaries of known to unknown areas. Using this technique to leverage data from perception sensors, rapid exploration can be achieved.

The frontier layer uses a tile-based approach to determine the most viable frontiers by taking the occupancy grid and processing it. When there are not enough viable frontiers, an expansive search of the map in a sweeping pattern radiating out from the UAS's location will occur. If frontiers are identified, the expansive search will stop and continuously update the existing frontiers. For a tile to be classified as a frontier, a configurable portion of the tile (e.g., 40%) must have transitional data (i.e., known-to-unknown). Once a frontier has been added to the frontier list, it will continually be updated until the transitional data drops under another configurable threshold (e.g., 25%). Once the frontier has enough data (i.e., it has been explored), it is added to a blacklist. The blacklist ensures that the tile will not be considered again when searching for new frontiers. This technique keeps the expansive search from taking too long as the size of the map increases. Figure 5 shows sample data from the world model frontier layer – areas shaded in green represent unexplored frontiers that are ripe for further exploration, while areas shaded in red represent full-explored or otherwise blocked areas of the operating environment.

The obstacle layer takes in Cartographer output and produces voxels of any nearby obstacles. It will primarily be used by a local planner for obstacle avoidance. The obstacle layer works by leveraging the voxel data produced by Cartographer and querying nearby voxels using a k-d tree search. A k-d tree is a space-partitioning data structure used for organizing points in a k-dimensional space. It is a binary search tree constrained to return the voxels around the UAS. The obstacle layer can be expanded to include various types of sensors as more sensors are added to the UAS payload. Figure 6 shows sample data from the world model obstacle layer.

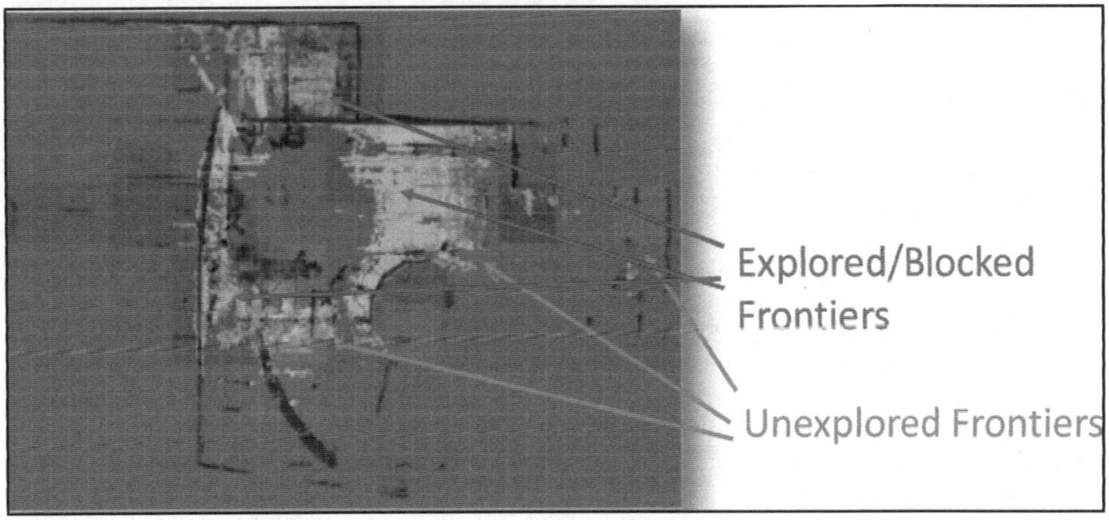

Figure 5. Sample Data from World Model Frontier Layer.

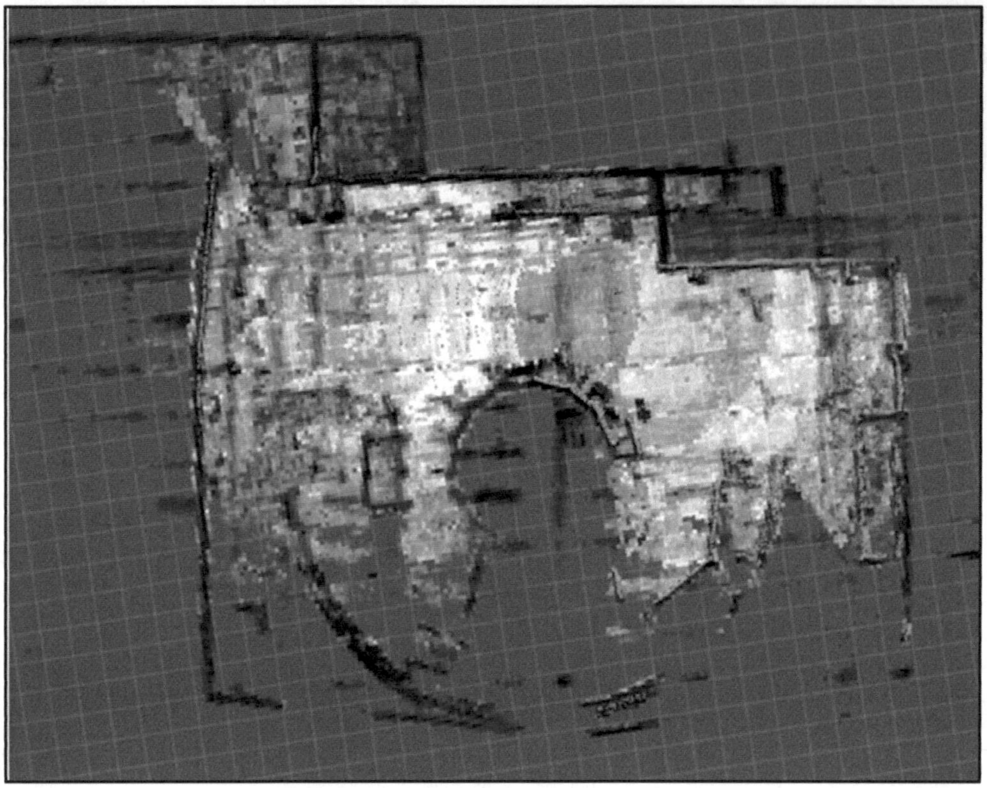

Figure 6. Sample Data from World Model Obstacle Layer.

NAVIGATION

Two planning algorithms that operate on information generated by Cartographer and the world model were implemented. The global planning algorithm, or exploratory planner, manages operation of the exploration behavior. It identifies which unexplored region to navigate to, tracks distance from the starting location (i.e., "home"), evaluates when to return home, and oversees the local planning process.

The optimal unexplored frontier is selected by considering the costs of the frontiers (the frontier with the lowest cost is considered the optimal frontier) which are composed of the following metrics:

- Euclidean distance from the UAS's position to the frontier
- UAS's maneuverability towards the frontier
- number of neighboring frontiers
- amount of transitional occupancy data in the frontier

Maneuverability is calculated by considering the velocity vector of the UAS in relation to the frontier position (i.e., a frontier's maneuverability cost is higher if it is behind the UAS's direction of flight). The exploratory planner uses voxels from the world model to determine if the frontier's location (centroid) is in free space. If the centroid voxels are occupied, then the exploratory planner begins searching the voxels vertically to see if there is an open space at a different altitude. If there isn't an open space, then the exploratory planner begins searching the voxels horizontally by incrementing based on the size of the UAS multiplied by a safety factor. The horizontal search begins at the altitude with the most open space. Once a sufficient open space is found, the position is sent to the local planner for path determination.

As the exploratory planner selects optimal frontiers and sets their locations, it is also monitoring the distance the vehicle has traveled from home. Using this distance and the maximum allowable speed of the UAS, an estimated time to return to home is calculated. The exploratory planner determines when to command the UAS to return home by subtracting this estimate from the allowable total flight time. The planner also monitors the remaining frontiers. If no new/unexplored frontiers exist, then the exploratory planner waits a configured amount of time to see if any are identified. The exploratory planner commands the UAS to return home if no additional frontiers are identified during the allotted time. Figure 7 shows sample frontier data with the prioritized frontier identified as a red dot and highlighted in yellow.

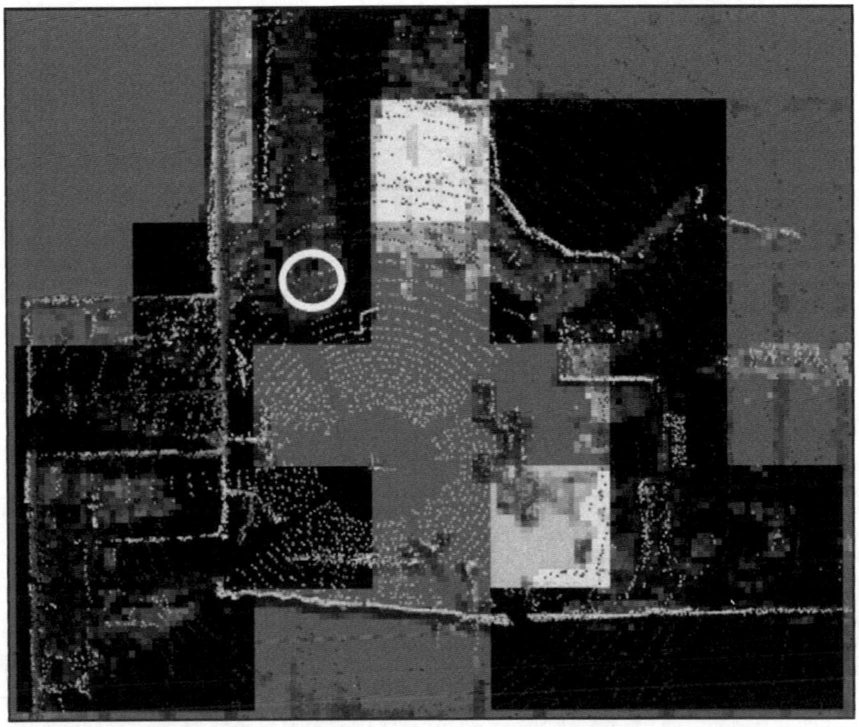

Figure 7. Frontier Layer with Prioritized Frontier.

A local planner employs an implementation of the rapidly exploring random tree star (RRT*) algorithm. The RRT implementation was designed to efficiently search non-convex, highly dimensioned spaces by randomly building a space-filling tree; however, RRT does not naturally provide optimized paths. Therefore, the RRT* algorithm extends RRT by examining the tree and reconnecting the nodes to create the shortest possible path to the goal position. This optimization allows the drone to navigate in free space and achieve the shortest path to the goal.

The first step in RRT* is to randomly sample free space from existing nodes in the tree. An unbiased random sampling results in uneven expansion into the free space which becomes exponentially worse as the number of nodes approaches infinity since the likelihood of randomly selecting a point of expansion in a previously explored area increases. One way to combat this is to bias the random sampling to expand from specific points in the tree. Our implementation creates "sweeping squares" that radiate outward from the start location to the goal. The bounds of the sweeping squares are used to guide the random samples to ensure even branching into free space.

A bias is applied to direct the pseudo-random sampling to the goal by calculating the Euclidean distance between the closest point and the goal.

Once new nodes are added to the tree, each node is checked for collisions against the free space voxels provided by OctoMap. New nodes must exist within the free space voxels to ensure that the tree does not begin to sample into unknown space. Sampling in unknown space could lead to unconstrained expansion which could invalidate any paths found within the tree. If the new node exists within a free space voxel, the closest existing node is found using the Euclidean distance. This closest existing node is then considered the parent of the new node. All neighbors of the new node within a configurable radius (two meters in our implementation) are also found to support the rewiring process. The cost for each of these neighboring nodes is calculated as the previous cost of its parent plus the Euclidean distance of the parent to the new node. This cost is used in the rewire process to optimize the path.

Once a parent of the new node is found, another collision check is performed to verify that the path connecting the two nodes in the tree is collision free (i.e., there are no occupied voxels within a certain distance). The collision check happens multiple times throughout the RRT* process and follows the same steps. A subset of the total voxels is used to speed up the processing time for the planner by first using a *k*-dimensional tree to query obstacle voxels within 10 meters of the search point. A normal vector is then calculated that connects the center of the voxel to the closest point on the path. A bounding sphere is "attached" to the intersection point between the voxel and path to ensure that no collisions will occur between obstacles and an inflated sphere around the drone. If this collision check fails, then the node is removed from the tree. As new nodes are added to the tree, existing nodes are evaluated with each new node. The proximal node is rewired to the new node if the total cost of the path decreases. By rewiring nodes in the tree, shorter paths may be found as new nodes are introduced. This eliminates the cubic nature of an RRT tree and introduces fanlike structures, shown in the orange circles in Figure 8b, as nodes are rewired.

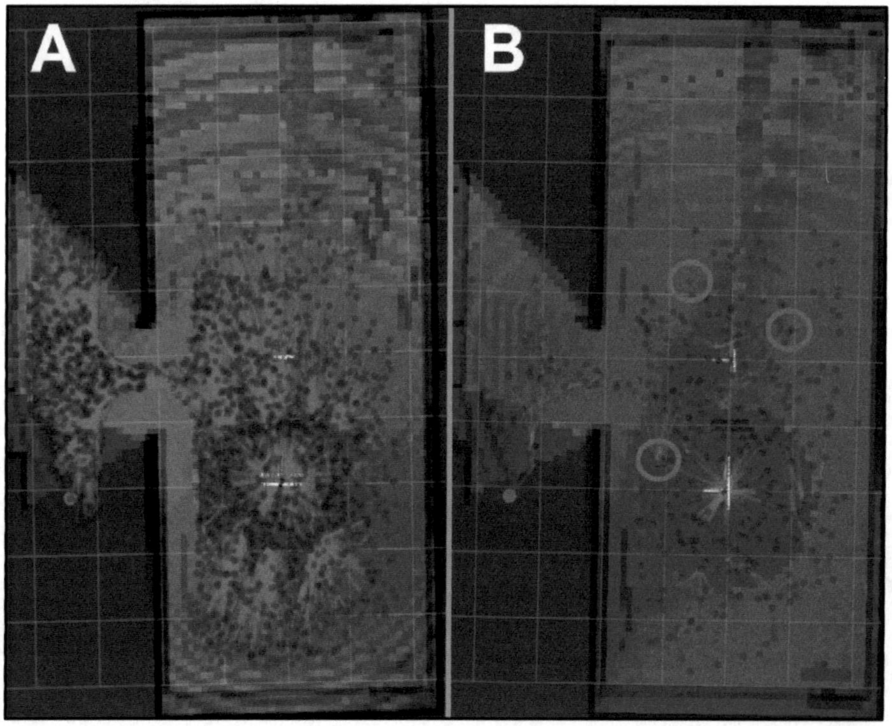

Figure 8. Sample Local RRT* a) tree without rewiring, b) tree with rewiring.

Finally, a custom path following controller (PFC) was developed and integrated to enable two key functionalities: the ability to point to a waypoint before moving and the ability to follow a line. The point-to-waypoint functionality is essential since our navigation sensors are located on the front side of the vehicle. While an Ouster lidar served as the primary perception sensor, an ultrasonic sensor with a limited horizontal and vertical field-of-view was mounted to the front of the UAS for the detection of glass objects, such as windows, that are problematic for lidar.

The ability to follow a line was added to the PFC during this project as the previous control scheme only attempted to traverse a waypoint and would not necessarily follow the straight line from its current position to the waypoint. RRT* plans are calculated to be collision free necessitating that the UAS follow a line exactly to the waypoint.

EXPERIMENTATION AND DEMONSTRATION

Testing was performed first in a simulated environment and then during live indoor flight tests at our facilities. Gazebo is a simulation software which supports simulated sensor data. The developed system was evaluated on "completeness" of exploration by comparing the ratio of mapped volume to the total volume of two test spaces: an approximately 25m x 18m confined and cluttered test fixture (see Figure 6 and Figure 7) and an approximately 82m x 60m space that was more open but generally bounded by objects (see Figure 3 and Figure 4). On average, approximately 91.5% of the test spaces were explored over ten test flights through each space.

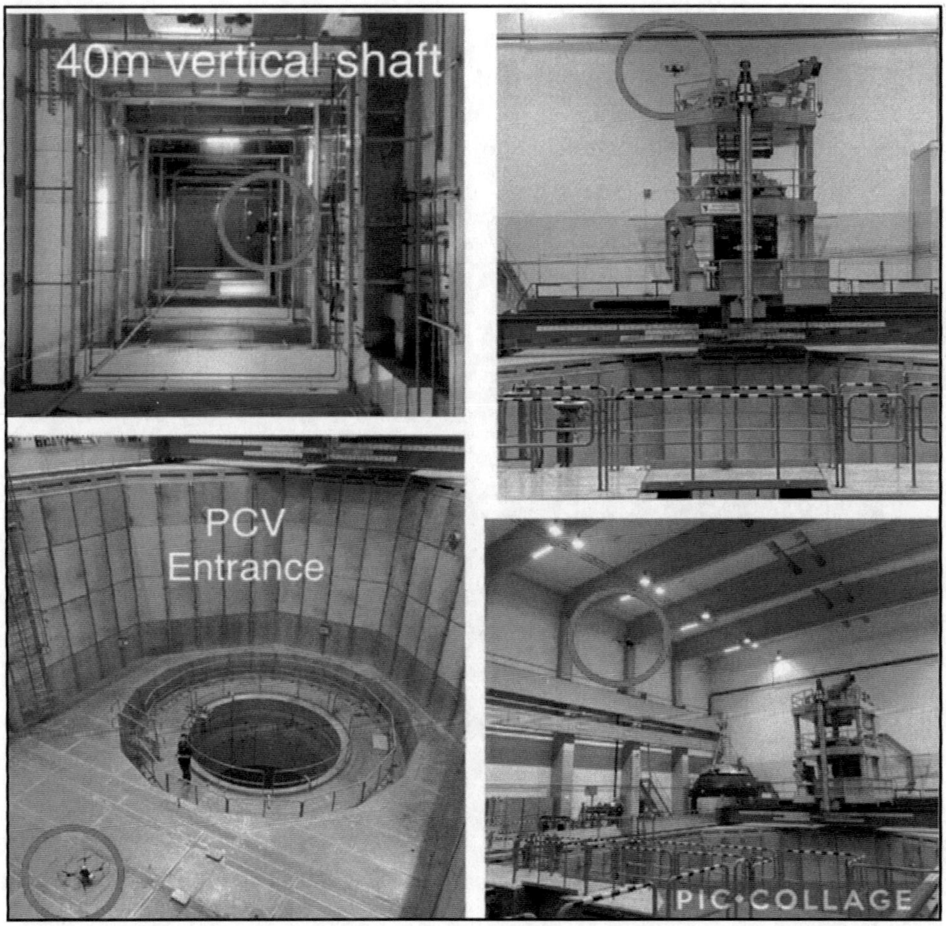

Figure 9. SwRI UAS Exploring AKW Zwentendorf Reactor Building.

Upon completion of development and testing, further experimentation occurred at the European Robotics Hackathon (EnRicH)* at the AKW Zwentendorf nuclear power plant outside of Vienna Austria in late 2021. EnRicH provided a venue to test and improve robotic systems in a simulated disaster scenario using a real-world environment. While the official event took place on pre-planned courses and environments, additional space and rooms were available for ad hoc testing. Figure 9 shows several images of the developed UAS exploring the large reactor room at the power plant. The image at upper left shows the UAS ascending and mapping a 40m vertical shaft, which would be used to raise fuel rods from ground level into the reactor room. The image at lower left shows the UAS navigating and mapping near where the primary containment vessel (PCV) would reside (and that would be flooded during operation). The images at right show the UAS exploring and mapping other areas within the reactor room. Figure 10 shows a bird's-eye-view of the post-processed map generated using data collected during one of the test runs. Several features of the reactor room are clearly visible in the map, including the fuel rod crane used to insert and remove fuel rods from the reactor, the PCV entrance, the staging tank where fuel rods are held and sub-merged before insertion into the reactor or after removal from the reactor, and the 40m vertical shaft used to raise fuel rods into the reactor room.

In addition to exploring and mapping the reactor room, we also tested the system by exploring and mapping the power plant's turbine room and cooling tower. Figure 11 shows another post-processed map of the turbine room with several distinct features visible, including the turbine blades.

After returning from the EnRicH event, we looked for additional opportunities to test and ex-periment with the developed UAS. We were able to use the UAS to explore and map two local underground caves near our headquarters in Texas. Figure 12 shows an image of our UAS navi-gating inside one of the caves and Figure 13 shows an x-ray view of a post-processed map of the cave.

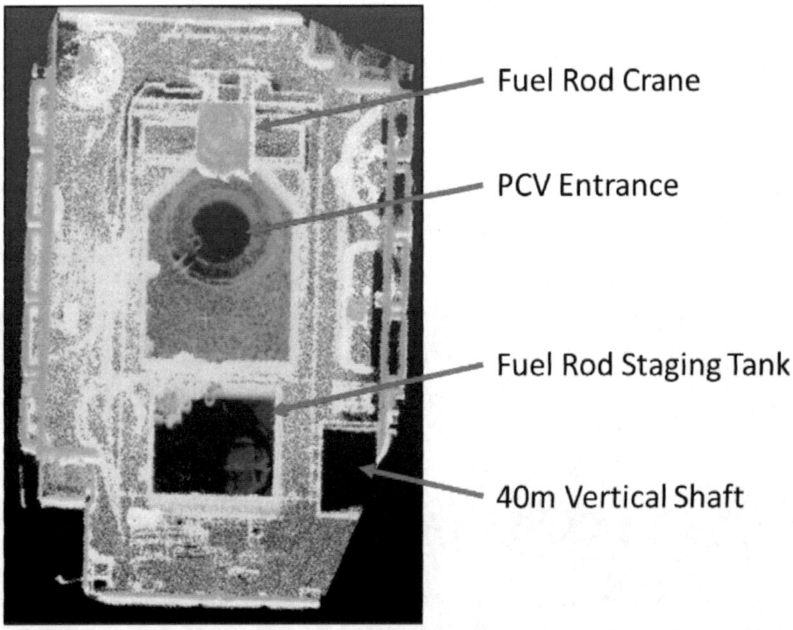

Figure 10. Bird's-eye-view of Post-processed Map of AKW Zwentendorf Reactor Building.

* https://enrich.european-robotics.eu/

Figure 11. Post-processed Map of AKW Zwentendorf Turbine Room.

Figure 12. SwRI UAS Exploring Local Texas Cave.

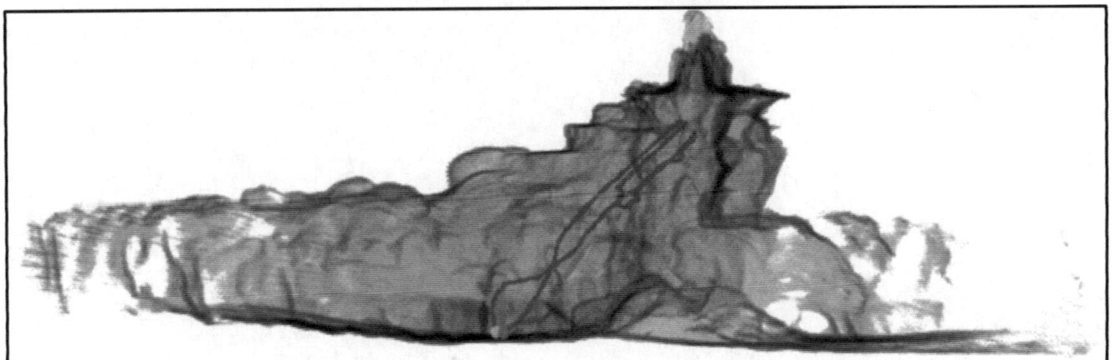

Figure 13. Post-processed Map of Cave.

CONCLUSION

This paper has summarized development and testing of a custom UAS platform and methods and algorithms to enable autonomous exploration of unknown and potentially GPS-denied environments. The system employs SLAM using the Google Cartographer open-source toolset to localize and map the environment. A custom modular world model framework was developed with two specific layers relevant to the exploration behavior: an obstacle layer to support navigation and obstacle avoidance, and a frontier layer to identify unexplored regions of the environment. A global frontier planner was implemented to evaluate and prioritize frontiers and a local planner utilizes RRT* to plan collision-free trajectories to the next frontier. Finally, a path following controller was implemented to generate linear and angular velocity control commands to follow the collision-free trajectories between frontiers.

We also summarized real-world experimentation in industrial buildings and natural underground caves and presented sample results of operation and high-fidelity post-processed maps. The experiments have demonstrated the feasibility and utility of the developed system in real-world environments and applications. Recent work has sought to explore other sensing modalities and optimize algorithms to operate on size, weight, and power (SWaP)-constrained platforms and hardware. Specifically, we have investigated using stereovision and other lighter range sensors to replace the heavy lidar utilized in this work and are evaluating the tradeoffs of reduced weight and power requirements versus the reduced range and resolution available with these sensors. Reducing the SWaP requirements would allow us to deploy the system on a smaller UAS and ultimately explore more confined spaces that the current platform is unable to reach.

CYBERWARFARE AND DRONES AUTONOMOUS ETHICAL HACKING

Laura Samsó Pericón*

How would you define Ethical Hacking? There are several definitions, such "Any hacking action towards the risk reduction of a situation to avoid bigger risks and challenges." Nefarious hackers are everyday threatening companies to expose data or shut down vital operations or critical infrastructures. On the other hand, sensors are becoming more and more connected and capable, exploiting what's possible, but at the same time exposing new vulnerabilities to hacking. The big leap in networking activities in the last years has accelerated a development where cyber security has to face new threats in exponentially growing numbers. Artificial Intelligence (AI) will play a major role in defining new threats lead by increasing automation of the systems of systems. New & more services will take networking in the 6th dimension populated with systems such as autonomous air taxis, drones, satellites, aircrafts for a new demanding society. I will present - Use cases: selected aspects where drone hacking, normally negatively tainted, can be performed for ethical causes especially in the cyberwarfare area.• Discuss how this can be done and illuminate technical aspects in the area of hacking drones• Describe the challenging environment were drones, networked sensors, AI and Internet of Things (IoT) exchange data & make decisions.

INTRODUCTION

Recently, the World Economic Forum (WEF) has released an intriguing analysis[1] on the future of jobs, presenting several key findings. Firstly, the report highlights the divergence of outcomes for labor markets worldwide in 2023 as a result of economic, health, and geopolitical trends. Secondly, the adoption of technology will continue to be a significant driver of business transformation in the next five years. Thirdly, environmental, technological, and economic trends will have the most considerable impact on job creation and destruction. Fourthly, the likelihood of technology adoption is highly evident in big data, cloud computing, and Artificial Intelligence (AI). Finally, the WEF expects the net impact of these technologies on jobs to be positive over the next five years, with cybersecurity and big data being key contributors, as well as humanoid and non-humanoid robots in the loop.

---

* Cyber R&D and Drones, www.laurasamso.com

It is evident from the analysis above that the long-discussed human-machine frontier is closer than ever, and we will eventually witness transhumanism as a regular human path that will significantly impact how society functions and lives.

The gradual takeover of many operations by automation is leading to a future of autonomous operations and services. It is crucial to distinguish between a trend and what is merely "trendy" when discussing technological development. A trend can be defined as something that society requires, catalyzed by a new technology, and it evolves across various areas of an ecosystem. Initially, it may be just isolated dots, but eventually, it converges onto the main path and stays with us[2].

In recent years, technology development has been accelerated due to unexpected events, commonly known as "black swans," which have changed society. These events have also altered the way wars and conflicts are fought, exposing vulnerabilities and increasing the rate and complexity level of attacks.

The rise in interconnected sensors, coupled with advancements in drones, self-driving cars, AI, 5G/6G, satellites, machine learning, automation, visualization, smart cities, blockchain networks, big data, the Internet of Things (IoT), the Internet of Senses, cloud, and quantum computing is laying down a catalytic foundation that makes a significant leap in operations possible. However, it also exposes them to new threats, opening the door to hackers. But do we have to consider hacking as overwhelmingly bad? How is it in the military environment?

The perspective on hacking's morality depends on the situation. In some cases, hacking could be beneficial in mitigating potential threats, introducing the concept of ethical hacking. For the purposes of this paper, we define hacking as any action taken to de-risk a situation and avoid significant damages, whether tangible or intangible. However, this action of de-risking could be perceived by adversaries as an act of war. Therefore the technological aspects becomes especially relevant in crisis or conflict, adding complexity to the definition of warfare in the age where technology is changing the way war is waged.

We are currently in an era where war is fought over cables and chips instead of battlefields. As technology continues its globalization across our physical landscape, boundaries are being eroded, and one of the significant challenges is defining what war is and when it should be declared and what means ethical in this environment.

The definition of war in modern times is more complex than ever before. With the rise of the internet and widespread access to knowledge, anyone with an internet connection can declare "war" against anyone else, whether they are human, government, or nation-state. The concept of war transcends traditional boundaries, making it a challenge to approach. This issue is not limited to the defense arena but also extends to the commercial and institutional fields, such as law enforcement or companies trying to protect their networked products/services.

In terms of defense, modern military doctrines describe the battlespace/conflict as being defined by five dimensions: land, sea, air, space, and cyberspace.

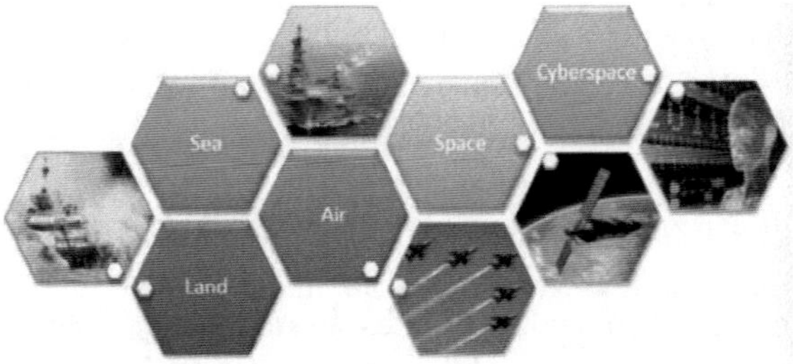

Figure 1. Battlespace dimensions[3]

The cyberspace dimension has become increasingly important, and governments and military organizations around the world have recognized its significance. As a result, they are developing and implementing strategies to protect their systems and operations from cyber threats. Traditional defense mechanisms are not always effective in this new domain, and new approaches are needed.

The Global Defense Cybersecurity market size is expected to reach $33.7 billion by 2028 (see Figure below).

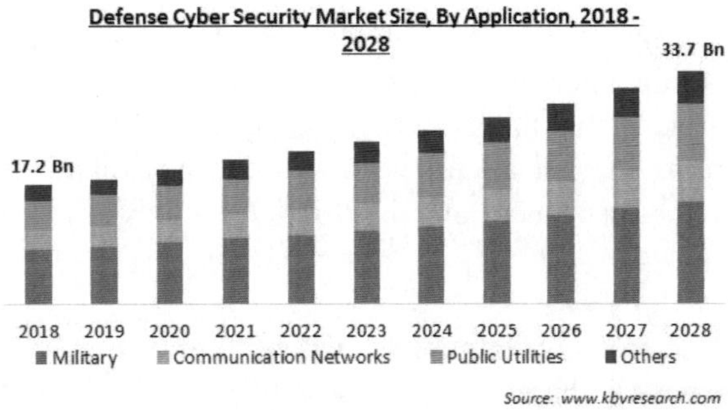

Figure 2. Cyber security market size by Application 2018-2028.

Cyberspace[3] is generally defined as "*domain characterized by the use of electronics and the electromagnetic spectrum to store, modify and exchange data via networked systems and associated physical infrastructures*".

The use of cyberspace in modern conflicts has created both communication capabilities and vulnerabilities for adversaries to exploit. The widespread availability of resources and anonymity in cyberspace makes it easy for enemies to inflict harm and adapt technology for military purposes. As a result, any conflict in the 21st century is likely to include cyber operations that will impact critical infrastructure (i.e. power plants, pipelines, hospitals), cars, medical devices, logistic chains and military systems, leaving them exposed or being used themselves as vectors of attack.

The Ukrainian crisis has demonstrated the different facets of hybrid warfare, combining conventional military forces with operations in the areas of information, cyber, and economics; and making deterrence a key strategic priority.

The concept of Network Centric Warfare (NCW) in the military refers to a system of systems where all components are interconnected through sophisticated networks. As technology advances, the use of autonomous AI to make decisions according to different assessments becomes more relevant. Therefore, the concept of autonomous hacking may need to be considered to de-risk a conflict, taking into account its implications.

The use of drone strikes in combating terrorist violence has been debated, with trends suggesting that the escalation of drone wars came in response to increases in terrorism or conflicts. The future may see fully interconnected sensors controlled by autonomous AI, leading to the need to consider when and whether autonomous ethical hacking can be used to de-risk conflicts.

In conclusion, the concept of war has evolved with the rise of technology, and the cyber domain is a critical aspect of modern warfare and defense. Addressing the challenges presented by cyberspace requires collaboration, innovation, and investment from both the public and private sectors.

NETWORKED SYSTEMS

As sensors and networking capabilities continue to advance, the military is shifting towards a Network Centric Warfare (NCW) approach[4]. This approach allows for faster and more efficient decision-making, but it also creates new vulnerabilities to hacking and cyber-attacks.

Similarly, in the consumer society, systems of systems such as cars, air taxis, telehealth services and other sensors are becoming increasingly connected and vulnerable to hacking. This creates new challenges for cybersecurity as anything with software in it can be considered a computer or a sensor vulnerable to attack. In critical infrastructure and natural disaster scenarios, drones are increasingly being used for assistance and distributed services, but they too are vulnerable to hacking.

In the military, battle networks or sometimes called "kill chains" are not new, they have been used for years to detect what is happening, process the data into information, decide course of actions, communicate, act on the decisions, and asses the effectiveness of the actions taken.

During WWII, with the use of new technologies such radar, sonar, radio comms and aerial reconnaissance, those networks also became a target. At the same time, the amount of information that the sensors produced, the speed of communications, and the magnitude of tactical advantage have changed and became more complex.

Currently the advantage lays in those who can see farther and clearer and act faster and in greater range; denying at the same time the other side the ability to do the same[5].

The modern battlefield has evolved to include a complex network of interconnected systems and processes. This network, also known as a battle network, is composed of five functional elements that work together to provide timely and effective responses to any threats or challenges. These elements combine processes, people, and technology enablers to ensure that the network operates efficiently and delivers the required capabilities.

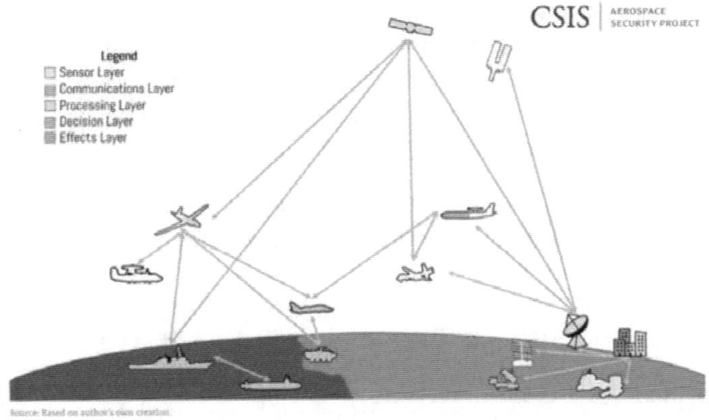

Figure 3. Diagram of Five functional elements of a Battle Network.

Former USAF Chief of Staff, General David L. Goldfein [4,6], has emphasized the importance of connected networks for future combat, stating that victory will depend less on individual capabilities and more on the integrated strengths of a connected network.

NATOs cyberwarfare doctrine[7] also takes into account cyber capabilities into multidomain operations in future high-intensity warfighting scenarios.

One example of the new concept of warfare is DARPA's "Mosaic Warfare"[8] which involves combining individual platforms to create a larger attack. The goal is to overwhelm the enemy with multiple weapons and sensor platforms from various domains, including ground, air, space, sea, and cyberspace.

Figure 4. DARPA Mosaic Warfare Concept[9]

To achieve this, all the information gathered by sensors across the network needs to be shared and processed in real-time, which requires a Multi-domain Command and Control System (MDC2S) with a combat cloud[10] at its core. The combat cloud is a manifestation of Network Centric Warfare (NCW) [11], and uses advanced digital networks and commercial cloud technologies to improve the efficiency, effectiveness, and resilience of air power. The tactical cloud must become an essential part of a future air combat system, but challenges such as cybersecurity, connectivity, interoperability, standards, and information sharing must be overcome.

The Future Combat Air System (FCAS) is an example of a system of systems concept led by a transnational group of European nations, including Spain, France, and Germany. A combat cloud is the driver for not only air power but also cross-domain synergy, according to retired Lieutenant General David Deptula. NATO is also working towards a new approach to combat operations that enables forces to plan and execute operations rapidly and effectively across all domains.

Finally, the use of AI, autonomy, and cyber warfare are essential components of modern battlefield synergy, and the paper will explore a use case and its technical challenges associated with this concept.

REVOLUTION IN WARFARE: ARTIFICIAL INTELLIGENCE

Artificial Intelligence (AI) is not a passing trend, but rather a new era of computing with the potential to transform societies. Although the roots of modern AI can be traced back hundreds of years to the work of philosophers, mathematicians, and scientists, there is no one definition that fully captures the concept. Broadly speaking, AI refers to systems that are able make autonomous decisions. The question of whether machines can truly be sentient and intelligent remains an open one.

Despite its long history, AI has recently experienced a quantum leap in its capabilities. Large Language Models (LLMs) are particularly noteworthy in this regard. These models are capable of generating their own training data to improve their performance and making decisions autonomously based on ecosystem inputs. Furthermore, they can fact-check themselves and confidently produce information, even when it is inaccurate or incorrect. As LLM architectures continue to evolve, the potential applications and impact of this technology will only continue to expand.

Venture capital[12] firms have invested billions of dollars into generative AI solutions over the last few years, and industries across the board will need to adapt to this disruptive technology. The democratization of AI is a reality that is here to stay, but it is also important that we ensure the technology is not misused or abused.

Some experts predict that we may be approaching the AI singularity, a point at which AI capabilities surpass human capabilities. According to futurist Ray Kurzweil[13,14] ,we may reach this point in 2030.

As we continue to develop AI technology, we must also grapple with ethical and security concerns. For instance, DARPA has launched an Artificial Intelligence Reinforcements (AIR) program with the goal of developing tools for achieving "dominant tactical autonomy." Similarly, the US Air Force's Skyborg project aims to use unmanned combat aerial vehicles (UCAVs) as unmanned wingmen to reduce costs and increase combat capabilities.

However, as with any powerful technology, there is the potential for nefarious uses of generative AI. Threat actors could use AI models to automate the development of new attack methods, or even create AI replicas of individuals, running in distributed mode, to hack into systems undetected; and how this could be escalated in a wider use case involving drone, battlefield or critical infrastructures and hackers. As we move forward, we must ensure that our systems are trained to make informed decisions in a safe and ethical manner.

AUTOMATION VS AUTONOMY

The integration of autonomy in military applications requires the implementation of regulations to ensure the safe operation of autonomous systems. Autonomy, in this context, refers to a system's ability to govern itself to accomplish a designated task based on its own situational awareness, planning, and decision-making capabilities .However, the deployment of autonomous weapon systems, also known as lethal autonomous weapon systems (LAWS), raises ethical concerns. Such systems can select and engage targets without further intervention from a human operator, prompting a need for clear distinction between automatic, automated, and autonomous systems. In this article, we explore the potential of autonomous ethical hacking capabilities to counter malicious attacks and mitigate risks.

The networked architecture of autonomous military as well as civil applications will also rely on unmanned vehicles on the ground. The development of self-driving cars is progressing towards achieving level 3 Conditional Driving Automation, according to the Society of American Engineers (SAE) classification system, with levels ranging from 0 to 5.

An autonomous weapon system (or lethal autonomous weapon system, LAWS) is a weapon system that, once activated, can select and engage targets without further intervention by a human operator. A distinction is often made by some between automatic, automated and autonomous systems, while others use these terms interchangeably.

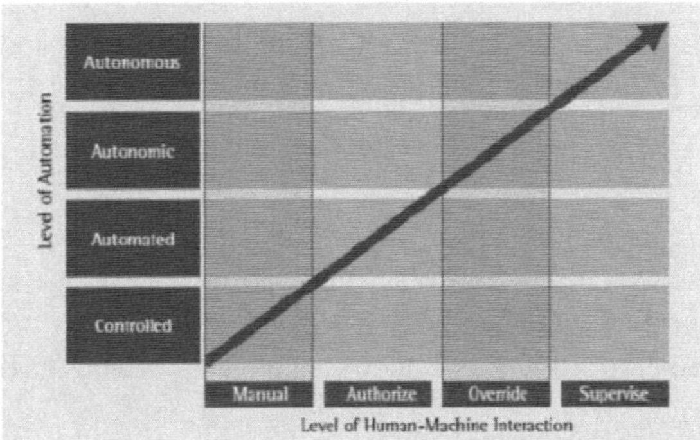

Figure 5. Relation between Level of Automation & Human-Machine Interaction[15]

For this article, we are looking at an autonomous ethical hacking capability in order to counter a malicious attack and to be able to de risk the whole mission.

Early this year the Pentagon released the update to DoD Directive 3000.09 "*Autonomy in Weapon Systems*" where states that autonomous and semi-autonomous weapon systems will be designed to allow commanders and operators to exercise appropriate levels of human judgment over the use of force. This connects to the question about the ethical aspect of military activities. A good example for that is the topic of collateral damage in drone strikes. Cyberwarfare as part of warfare cannot be without rules including the ethical aspect. International Humanitarian Law (IHL) continues to be applicable at all time, also for cyberwarfare.

CYBERWARFARE: THE NEW/MODERN BATTLEFIELD SYNERGY

We are currently experiencing a global shift in regulations and societal norms due to the emergence of a multipolar global order.

The use of unmanned aerial vehicles (UAVs), or drones, is expected to increase as technology advances and becomes more affordable. However, this also leads to a growing interest in cyber-attacks targeting both individual drones and entire UAV networks. Recent conflicts in countries such as Ukraine, Azerbaijan, and Syria have become testing grounds for various technologies, including autonomous and semi-autonomous drones, cyberwarfare, satellite communications, artificial intelligence, and robotics.

In 2016, the NATO Summit in Warsaw NATO Council Summit Warsaw[16], recognized cyberspace as a domain of operations that requires effective defense. Cyberwarfare presents new areas of ambiguity, such as the violation of third-party sovereignty, use of cyber-attacks by non-state actors, attacks on critical infrastructure, and the use of networks of sensors, robots, and drones, among others. These activities could be interpreted as an act of war and need to be evaluated accordingly.

A cyberattack is an electronic attack that aims to disrupt or corrupt a system, enterprise, infrastructure, or individual, where the assets may be digital, digital services, or physical assets with a cyber component. These attacks typically seek to compromise the confidentiality, integrity, and availability (CIA) of these assets. In Europe, legislative and policy frameworks, such as the Network and Information Security (NIS) Directive, the EU Cybersecurity Act (CSA), the Digital Single Market Strategy, and the proposed EU Cyber Solidarity Act, have been developed to combat cyber threats.[17]

Every nation is now writing and developing frameworks and eco systems in order to regulate and properly prepare for cyber-attacks.

The US Department of Defense (DoD) prepared a list of drones recommended for use by the US government in August 2020, known as Blue UAS, which aimed to prototype, scale, and secure commercial UAS technology for the DoD. Two years later, the Association for Unmanned Vehicle Systems International (AUVSI) and Fortress Information Security launched the first product of their Trusted Cyber Program, Green UAS, which addresses corporate cyber hygiene, product and device security, remote operations and connectivity, and supply chain risk management.

UAV cybersecurity programs are necessary because the current mission execution capability is still limited and is supposed to enhance the capabilities with the coming AI advancements.

Now after covering all the components the paper goes through identifying common drone challenges and vulnerabilities in preparation for the use case analysis.

COMMON VULNERABILITIES AND CHALLENGES

The process of a general hacking attack typically involves several phases, starting with the Planning and Reconnaissance phase, during which information is gathered about the target. This is followed by the Scanning/Weaponization phase, in which attack methods and vulnerabilities are identified and selected. The hacker then establishes the first point of contact with the target (Delivery) and proceeds to take over the system (Execution). To avoid detection and maintain access, the hacker may persist in their attack.

Given the capabilities of modern drones, these unmanned systems could potentially be used in an autonomous hacking event targeting a network or system of systems. Such an attack could leverage the advantages of drones, such as their mobility and ability to evade detection, to gain access to vulnerable systems and carry out a cyberattack. This underscores the need for robust cybersecurity measures and countermeasures to protect against such threats.

Figure below show classical high level UAS configuration with an air segment and a ground segment, each one with their respective elements.

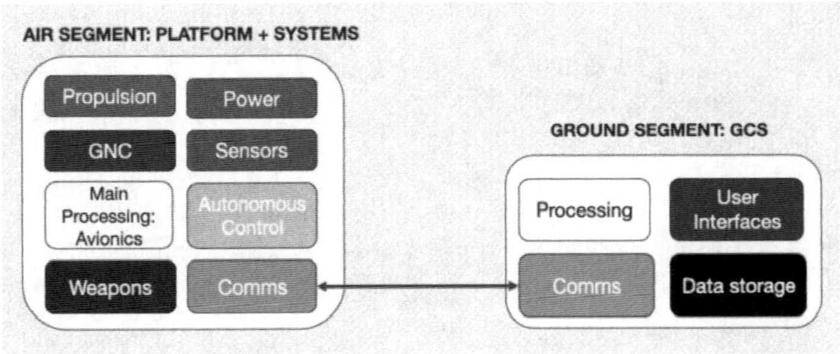

Figure 6. UAS configuration: Air segment + Ground segment.

Having a closer look to the air segment, Figure below presents some common vulnerable points such payloads, navigation and positioning systems (inertial system (INS), Global Positioning System (GPS)), radar to name a few.

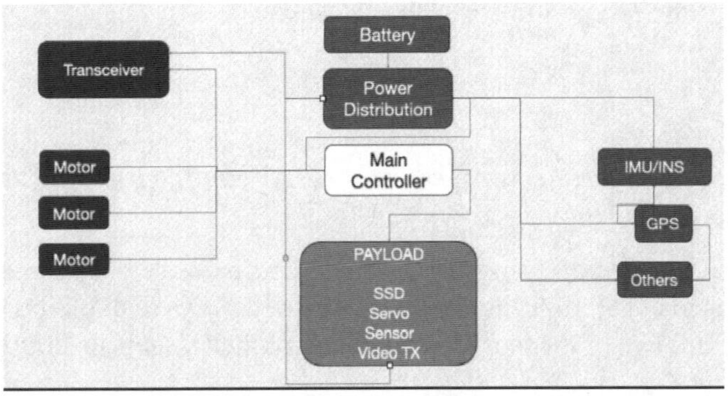

Figure 7. Generic UAV system architecture.

The main controller is the brain of the drone and controls all systems such navigation and positioning, the motors, the power distribution and the different sensors.

There are different types of attacks such application logic attacks and control system attacks.

Control System Attack could be hardware or software, for example to prevent H/W or CPU from behaving as programmed resulting in buffer overflow exploits and forced resets to load malicious code.

Application Logic Attack consists in altering data to the Control System, manipulation of sensors or the environment to provide false data, sensor data manipulation (waypoint attack), vehicle/ component state manipulation, NavData manipulation

Attacks to the payload/C2 data could results in data sensor stolen or replaced by fake video imagery: direct commands to the gimbal, a direct payload attack could result in the temporary or permanent damage to it.

The figure below presents an example for a high level UAS networking architecture and its different elements that could be hacked.

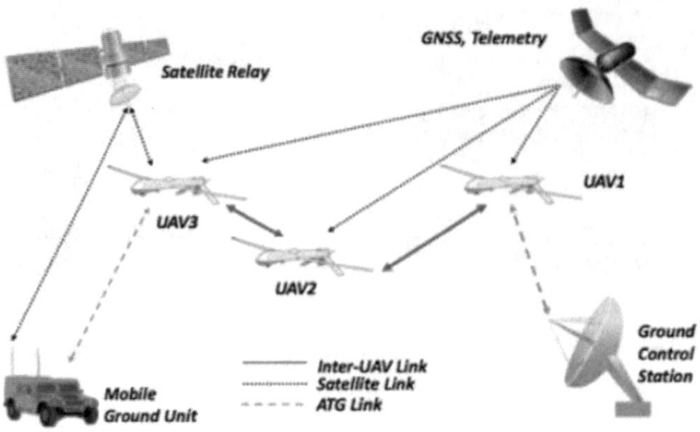

Figure 8. High level UAS Networking architecture[18]

Now we have covered all the components and fields that conform the scope of this paper so we are ready to go through the last part to cover the use cases

USE CASES

The environment were drones, networked sensors, AI and Internet of Things (IoT) exchange data & make decisions is challenging.

With the rise of the Internet of Things (IoT), more and more devices can connect online. This trend has led to increased cybercrime, as criminals find new ways to exploit these devices for their own gain. IoT and OT are two of the most commonly exploited targets in industrial control systems and critical infrastructure attacks.

The main risks posed by IoT hacking include theft of data, tampering with data, Distributed Denial of Service (DDoS) attacks and spying.[19]

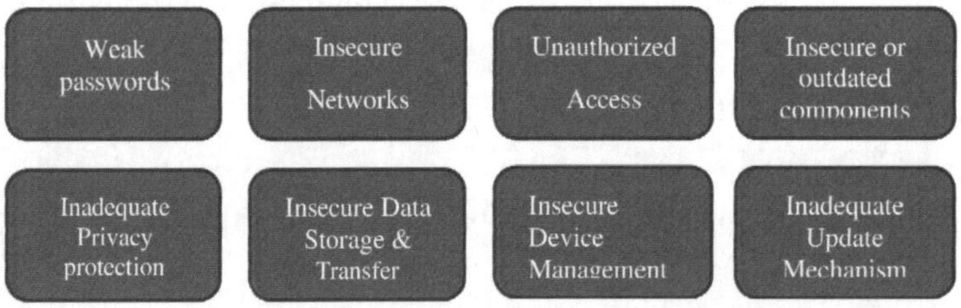

Figure 9. Common IoT vulnerabilities.

OT refers to the systems that control industrial processes and other critical infrastructure. These systems are often connected to the internet, making them a target for hackers.

The main risks posed by OT hacking include: Damage to equipment, Data theft, Hijacking of devices and Sabotage. In the image below we could observe different IoT related applications. The list is not exhaustive.

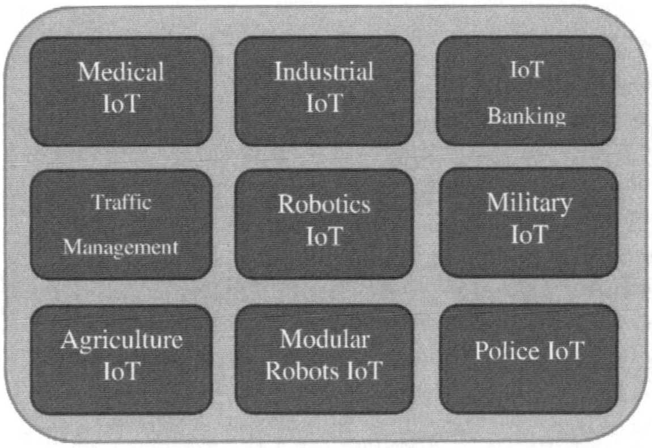

Figure 110. IoT application.

Let´s take the concrete use case where there are different layered components: UAS with IoT capability, that will be interfacing with different other layers such space, ground and sea.

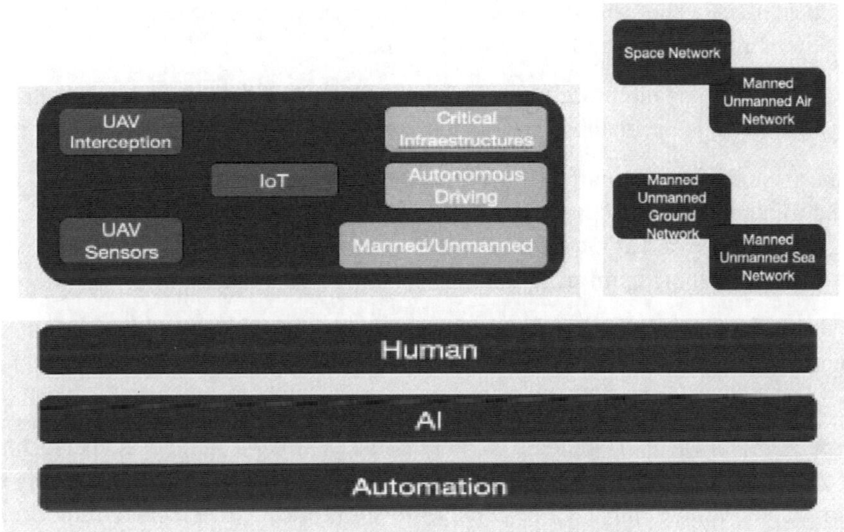

Figure 11. Layered components.

In addition, AI will handle the different mission features, at the same time that several cybersecurity measures. The image below shows a high level scenario involving all those layered components in a little bit more of detail.

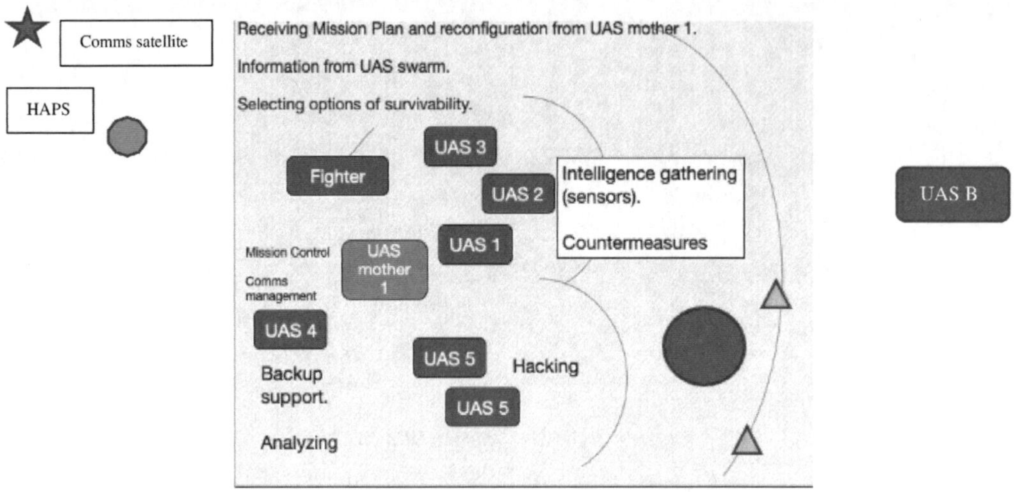

Figure 12. UAS use case scenario.

The scenario (Figure above) shows a critical infrastructure (red circle) and its ground elements (yellow triangle) located in a the country A. UAS team 1 belongs to country A. UAS team is composed by a mothership, a fighter and a swarm of drones.

This is the biggest energy supplier of country A and UAS team 1 detected an strange activity in the nearby: a drone (UAS B) is approaching. UAS team 1 does not know if UAS B is friend or foe.

Elements in the ground from country A are connected through IoT networks and could be hacked and/or used to hack other systems through the network. Moreover, the facility is occupied by several hundreds and the evacuation on time is not possible. Meanwhile the swarm of drones are assessing the situation gathering information.

UAS mother 1 sends several intents for communications to UAS B. None of them have response. This could be due to different reasons: comms are not working or UAS B does not reply for other unknown reasons.

In that situation, UAS swarm from team A begins assessing the situation, performs a risk analysis for the counter measures and re-plans the mission, if applicable. Networked AI evaluates all available information and there are two options: a) to shoot down the approaching drone or b) to hack the approaching drone in order to de risk the situation.

This hacking action could be interpreted also as act of war by the other country but, it was decided to do so because the potential risk for this interpretation would be reduced. This hack would represent an example for ethical application of hacking technology.

An increase in complexity assuming in the scenario above could be the following:

- A chatbot began to impersonate a commander, sending orders to the UAS team A mothership and replicating itself and disguise as a friend,
- In case of a swarm attack:
  - to use a node of team B as if it was team A for communication cell in the air and control the critical comms (or take over HAPS and disrupt personnel in the battlefield),
  - to hack communications while one of the swarm drones is assisting an emergency surgery in the battlefield,
  - to give UAS team B control over network A to use the shared capabilities under a false flag,
- to steal video streaming or to alter it in real time,

- to hack comms satellite that UAS team A is using,
- to hack any of the onboard UAS sensors,
- to hack and disrupt the air traffic over the scenario area,
- to hack the ground vehicles through the comms network (5G) with safety of life consequences.

These are some examples of possible hacking where the nefarious agents could take control of the drone where a decision about ethical implication would be made.

In the future the AI may detect the intrusion, assess the situation, chose the "right" response and replan in real time. The decision making will be decentralized and the execution too in order to enable a quick operational pace, introducing the ethical topic in the AI and regulatory development.

Autonomous ethical hacking has its pros and cons, and also should have its limits.

ETHICAL GOVERNANCE

In recent years, the adoption of digitalization has become prevalent across various markets, including the military and civil sectors, as it offers a means to reduce costs and enhance performance. In response, different nations have implemented rules and regulations related to drone operations and privacy, and are currently seeking to establish a framework for cyber security.

However, the act of hacking can result in collateral damage, and the responsibility for such actions lies with those who engage in them, particularly in the case of economic damages and prevention of cyberattacks.

One of the primary concerns associated with drone technology is the potential threat of civilian casualties, as well as adherence to international humanitarian law principles of proportionality and distinction, and the possibility of overuse.

Moreover, the lack of clear legal provisions pertaining to the prevention of cyber-attacks and liability for economic damages resulting from a drone-related cyber-attack may impact consumer and entity perceptions of drone usability in different sectors of the economy.[20]

CONCLUSIONS

Technology development is speeding up certain societal changes and also the way wars are fought. AI is widely accelerating changes that society would only have cop with in several years to now and adding to the discussions about the ethical dimension of autonomy of drones, in the civil and military areas.

ACKNOWLEDGMENTS

Thank you for all the people that supported me to produce this paper.

REFERENCES

1 World Economic Forum, "The Future of Jobs Report 2023", Insight Report, May 2023.

2 Webb, A., "The Signals are Talking", 2017.

3 Chairman of the Joint Chiefs of Staff, "The National Military Strategy for Cyber Ops 2006" , 2006.

4 NATO JAPCC, "Future Battlespace Management", Ed. 31, 2021.

[5] Harrison, T., "Battle Network and Future Force", CSIS Brief, 2021.

[6] NATO JAPCC, "Cybersecurity Challenges", Ed. 31, 2021.

[7] NATO, "Allied Joint Doctrine for Cyberspace Operations", 2020.

[8] CSBA, "Mosaic Warfare, Exploiting AI and Autonomous systems to implement decision-centric operations", 2020.

[9] DARPA, "Mosaic Warfare", www.darpa.mil

[10] NATO JAPCC, "Multi domain combat cloud", 2021.

[11] World Economic Forum, "Towards actionable governance on trustworthy AI", 2023.

[12] Kurzweil, R., "The Singularity is Near: When Humans Transcend Biology", 2006.

[13] Finantial Times, "It's all about the chips in the AI War", 2023.

[14] IEEE, "A Generative AI for Heterogeneous Network-on-Chip Design Space Pruning", 2022.

[15] NATO JAPCC, "Future Unmanned systems Technologies", 2017.

[16] NATO JAPCC, "Warsaw Summit Priorities", 2016.

[17] EU Commission, "Cyber: towards stronger EU capabilities for effective operational cooperation, solidarity and resilience", 2023

[18] Bezahdan, V., "Cyber-Physical Attacks on UAS Networks- Challenges and Open Research Problems", 2017.

[19] ECCouncil, "How Ethical Hackers Can Defend Against IoT and OT Hacking", 2021.

[20] VVAA, "Legal, Economic and Cyber Security Framework Considerations for Drone Usage", 2022.

AUTHOR INDEX